International Pocket Guide for HACCP

For All Food Industries

(Employees & Employers)

Jahangir Asadi

Vancouver, BC CANADA

Published by: Top Ten Award International Network
Vancouver, BC **CANADA**
Email: Info@TopTenAward.net
www.TopTenAward.net

Ordering Information:
Quantity sales. Special discounts are available on quantity purchases by universities, schools, corporations, associations, and others. For details, contact the "Sales Department" at the above mentioned email address.

International Pocket Guide for HACCP/J.Asadi—1st ed.

ISBN: 978-1-7775268-5-6

Contents

This book is dedicated to my first manager, **Mir Ahmad Sadat***, a Distinguished man of ISIRI, who has been my supporter and has encouraged me greatly and believed genuinely in my ability with honesty and self-confidence.*

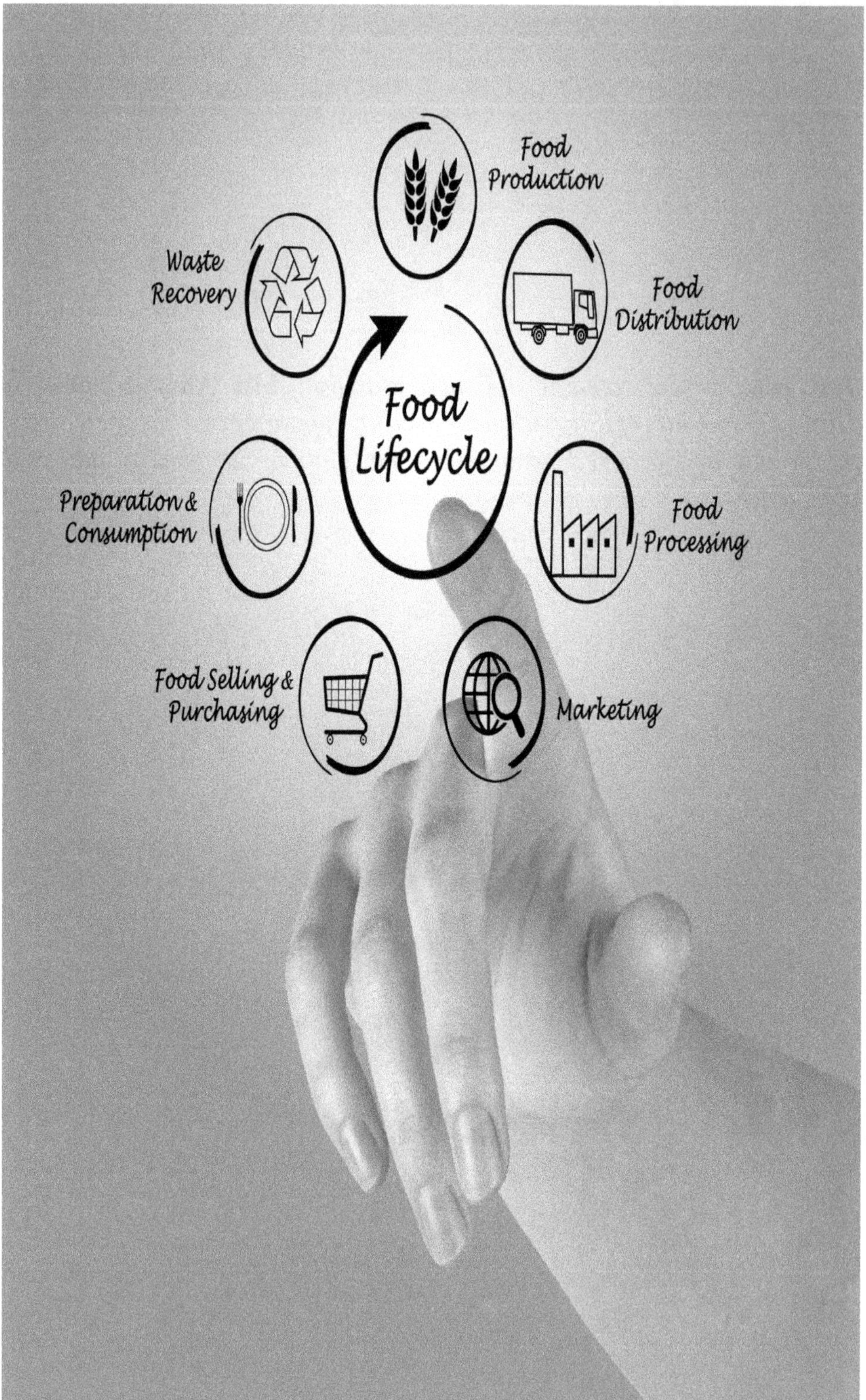
Food Production
Food Distribution
Food Processing
Marketing
Food Selling & Purchasing
Preparation & Consumption
Waste Recovery
Food Lifecycle

Top Ten Award International Network

Top Ten Award international Network (TTAIN) was established in 2012 to recognize outstanding individuals, groups, companies, organizations representing the best in the public works profession.
TTAIN publishing books related to international Eco-labeling plans to increase public knowledge in purchasing based on the environmental impacts of products.
Top Ten Award International Network provides A to Z book publishing services and distribution to over 39,000 booksellers worldwide, including Apple, Amazon, Barnes & Noble, Indigo, Google Play Books, and many more.
Our services including: editing, design, distribution, marketing
TTAIN Book publishing are in the following categories:

Student
Standard
Business
Professional
Honorary

We focus on quality, environmental & food safety management systems , as well as environmnetal sustain for future kids. TTAIN also provide complete consulting services for QMS, EMS, FSMS, HACCP and Ecolabeling based on international standards.

The Hazard Analysis Critical Control Point system is a preventative system for assuring the safe production of food products. It is based on a common sense application of technical and scientifi c principles to a food production process.

TTAIN has enough experiences to help create new ecolabeling programmes in different countries all over the world. For more detail visit our website : http://toptenaward.net and/or send your enquiery to the following email:

info@toptenaward.net

CHAPTER 1

About HACCP

The Hazard Analysis Critical Control Point system is a preventative system for assuring the safe production of food products. It is based on a common sense application of technical and scientific principles to a food production process.

The most basic concept underlying HACCP is that of prevention. The food processor/handler should have sufficient information concerning the food and the related procedures they are using, so they will be able to identify where a food safety problem may occur and how it will occur. If the 'where' and 'how' are known, prevention becomes easy and obvious, and finished product inspection and testing becomes needless. The HACCP program deals with control of factors affecting the ingredients, product and process. The objective is to make the product safely, and be able to prove that the product has been made safely. The where and how are the HA (Hazard Analysis) part of HACCP. The proof of the control of the processes and conditions is the CCP (Critical Control Point) part. Flowing from this basic concept, HACCP is simply a methodical and systematic application of the appropriate science and technology to plan, control and document the safe production of foods. HACCP is not the only method in ensuring that safe food products are manufactured. The plan will be successful when other procedures are in place such as sanitation standard operating procedures (SSOP's) and by using good manufacturing practices (GMP's). These programs are fundamental in the development of a successful HACCP plan. SSOP's should include personal hygiene practices as well as daily sanitation of the food contact surfaces and equipment. Good sanitation practices are the foundation of manufacturing and preparing safe food. HACCP is a management system in which food safety is addressed through the analysis and control of biological, chemical, and physical hazards from raw material production, procurement and handling, to manufacturing, distribution, and consumption of the finished product. For successful implementation of an HACCP plan, management must be strongly committed to the HACCP concept. A firm committed to HACCP by top management, provides company employees with the sense of importance of producing safe food.

CHAPTER 2

Definitions

CP Decision Tree: A sequence of questions to assist in determining whether a control point is a CCP.

Continuous Monitoring: Uninterrupted collection and recording of data such as temperature on a stripchart, or a continuous recording thermometer.
Control: (a) To manage the conditions of an operation to maintain compliance with established criteria. (b) The state where correct procedures are being followed and criteria are being met.

Control Measure: Any action or activity that can be used to prevent, eliminate or reduce a significant hazard.

Control Point: Any step at which biological, chemical, or physical factors can be controlled.

Corrective Action: Procedures followed when a deviation occurs.

Criterion: A requirement on which a judgment or decision can be based.

Critical Control Point (CCP): A point, step or procedure at which control can be applied and is essential to prevent or eliminate a food safety hazard, or reduce it to an acceptable level.

Critical Defect: A deviation at a CCP which may result in a hazard.

Critical Limit: A maximum and/or minimum value to which a biological, chemical or physical parameter must be controlled at a CCP to prevent, eliminate or reduce to an acceptable level the occurrence of a food safety hazard.

Deviation: Failure to meet a critical limit.

Document: a written or printed paper that gives information about or proof of something and/or a computer file containing data entered by a user.

HACCP: A systematic approach to identification, evaluation, and control of food safety hazards.

HACCP Plan: The written document which is based upon the principles of HACCP and which delineates the procedures to be followed to assure the control of specifi c process or procedure.

HACCP System: The result of the implementation of the HACCP Plan procedures to be followed.

HACCP Team: The group of people who are responsible for developing, implementing and maintaining the HACCP system.

Hazard: A biological, chemical, or physical agent that is reasonably likely to cause a food to be unsafe for consumption.

Hazard Analysis: The process of collecting and evaluating information on hazards associated with the food under consideration to decide which are signifi cant and must be addressed in the HACCP plan.

Monitor: To conduct a planned sequence of observations or measurements to assess whether a CCP is under control and to produce an accurate record for future use in verifi cation. Prerequisite Programs: Procedures, including Good Manufacturing Practices, that address operational conditions providing the foundation for the HACCP system.

Preventative Measure: Physical, chemical, or other factors that can be used to control an identified health hazard.
Sensitive Ingredient: An ingredient known to have been associated with a hazard for which there is a reason for concern.

Severity: The seriousness of the effect(s) of a hazard.

Step: A point, procedure, operation or stage in the food system from primarily production to final consumption.

Validation: That element of verification focused on collecting and evaluating scientific and technical information to determine if the HACCP plan, when properly implemented, will effectively control the hazards.

Verification: Those activities such as methods, procedures, or tests in addition to monitoring, that determines if the HACCP system is in compliance with the HACCP plan and/or whether the HACCP plan needs modifi cation and revalidation.

CHAPTER 3

HACCP Principles

1 Conduct a Hazard Analysis

2 Identify Critical Control Points (CCP's)

3 Establish Critical Limits for Each CCP

4 Establish CCP Monitoring Procedures

5 Establish Corrective Actions

6 Establish Record keeping Procedures

7 Establish Verification Procedures

Principle 1: Conduct a Hazard Analysis

The hazard analysis looks at different factors that could affect the safety of your product. This analysis is done for each step in your production process. It's important to remember that you are dealing with safety, not quality issues.

The hazard analysis is actually completed in two stages. The first stage identifies food safety hazards that are present in your process. The second stage evaluates these food safety hazards as to whether they are "reasonably likely to occur." If the HACCP team decides that a food safety hazard is likely to occur, then they need to fi nd and list any preventive measures that could be used to control those food safety hazards. Preventive measures are defined as: "Physical, chemical, or other means that can be used to control an identifi ed food safety hazard."

INGREDIENT RELATED HAZARDS: As you evaluate the hazards in your process, don't forget about ingredient related hazards. Everything that goes into your product needs to be evaluated. Ingredient specifications, provided by your supplier, should give you details on the materials/ingredients being sold, including statements that the materials/ingredients are of food grade and are free of harmful components.

For example, the ingredient specification for dried legumes (beans) might state that there will be fewer that 5 small rocks or stones per ten pound bag and that no harmful pesticides were used in the growing process.

Principle 2: Identify Critical Control Points (CCP's)

A critical control point is defi ned as "A point, step or procedure in a food process at which control can be applied and, as a result, a food safety hazard can be prevented, eliminated, or reduced to acceptable levels.
The HACCP team uses the list of food safety hazards and preventative measures they developed during the previous hazard analysis step to determine their critical control points.
CCP's may include, but are not limited to:
- Chilling or freezing
- Cooking
- Certain processing procedures; smoking, curing, acidification

Steps that are CCP's in one facility may or may not be CCP's in your facility.
When making a HACCP plan,
each facility must look at the unique conditions present in that facility.

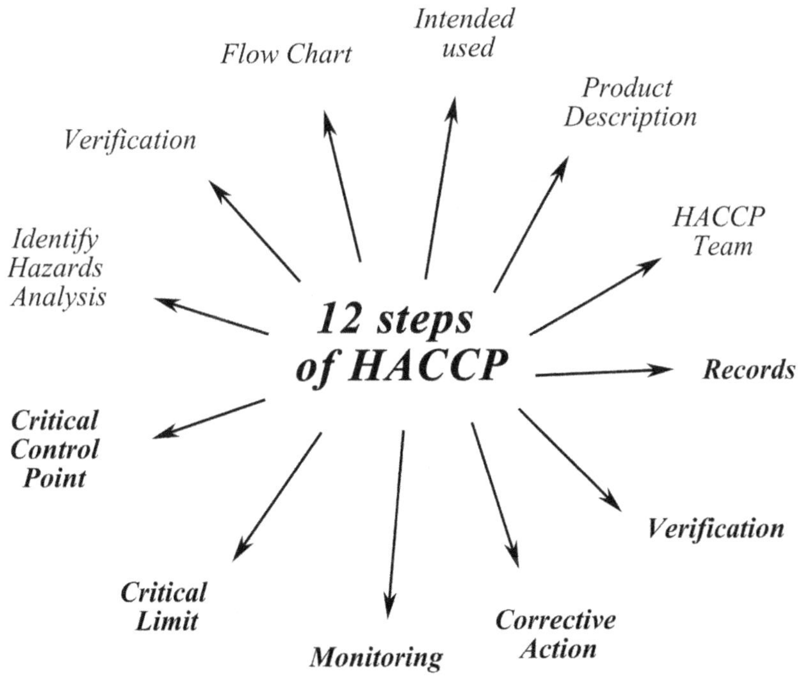

Principle 3: Establish Critical Limits for Each CCP

A critical limit is defined as "The maximum or minimum value to which a physical, biological, or chemical hazard must be controlled at a critical control point to prevent, eliminate, or reduce to an acceptable level the occurrence of the identified food safety hazard." Critical limits serve as boundaries of safety for each CCP.

Often they are a numerical value (whether that is temperature, pH, etc.) that must be reached to assure that a food safety hazard has been controlled.

[A note about Critical Limits -- When your HACCP team establishes critical limits for your specific facility, know that those limits may never be less strict than the current regulatory standards.]

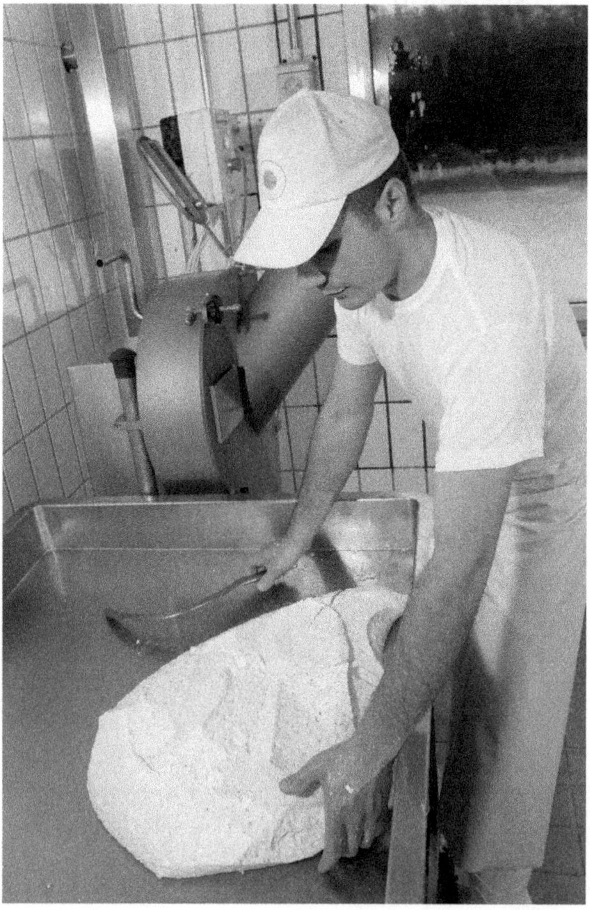

Principle 4: Establish CCP Monitoring Procedures

Monitoring is a fundamental part of any HACCP system. It consists of observations or measurements that check to see that your CCP's are operating under control. Monitoring serves three main purposes:

First,

it tells you when there's a problem at a CCP, and control has been temporarily lost. (This allows you to take corrective actions right away.)

Second,

it tracks the system's operation and can help identify dangerous trends that could lead to a loss of control. (This allows you to take preventive action to bring the process back into control before it goes beyond the critical limits.)

Third,

it provides written documentation of your compliance with the HACCP regulation. (This information can be used to confirm that your HACCP plan is in place and working right.)

For each CCP the HACCP team will need to define the monitoring procedure and its frequency (hourly, daily, weekly, etc.) that best tracks that CCP. It's also important to thoroughly train the employee(s) that will be responsible for each monitoring procedure and frequency.

Monitoring Requires Precision

Monitoring a CCP is a big responsibility. Employees must be properly trained and need to understand the reasons for careful monitoring procedures. Specify in your monitoring procedures, every important detail about...

- Who will do the monitoring
- What is being monitored
- When it is done, and
- How it is done

For example, when taking the temperature of a piece of meat, be specific as to where you took the temperature. Remember that all records and documents associated with a CCP's monitoring should be dated and signed or initialed by the person doing the monitoring and the results recorded.

Principle 5: Establish Corrective Actions

Corrective actions are defined as "Procedures to be followed when a deviation occurs." A deviation is defined as a "failure to meet a critical limit." Corrective actions are taken when monitoring shows you that a food safety hazard has gotten out of control at a CCP.
The best way to handle deviations is to have a plan of action already in place. In general, corrective action plans are used for:

1. Determining the disposition of non-complying product;
2. Correcting the cause of the non-compliance to prevent a reccurrence:
3. Demonstrating that the CCP is once again under control (this means examining the process or product again at the CCP and getting results that are within the critical limits).

As with the monitoring procedures, specific corrective action procedures must be developed for each CCP.

Principle 6: Establish Recordkeeping Procedures

Record keeping procedures are important in making and keeping an HACCP system effective. Every time monitoring procedures are done, corrective actions are taken, or production equipment is serviced, a detailed record of that activity is made. This continual recording of this information allows you to keep track of everything that goes on in your facility.

You can think of HACCP records in two ways, development forms and day-to-day "working" logs. The development forms are all of the supporting documentation that go into building your first HACCP plan. The "working" logs are the sheets of paper where you collect the details of what happen on the production floor.

You may wish to use the example forms located in Appendix I, or you may wish to create your own forms.

Generally, the records kept in the total HACCP system include the following:

• The HACCP plan itself and all supporting documentation.

• Records (including product codes) documenting the day-to-day functioning of the HACCP system such as daily monitoring logs, deviation/corrective action logs, and verification logs.

Principle 7: Establish Verification Procedures

Every establishment should validate the HACCP plan's adequacy in controlling the food safety hazards identified during the hazard analysis, and should verify that the plan is being effectively implemented.

1. **Initial validation.** Upon completion of the hazard analysis and development of the HACCP plan, the establishment shall conduct activities designed to determine that the HACCP plan is functioning as intended. During this HACCP plan validation period, the establishment shall repeatedly test the adequacy of the CCP'S, critical limits, monitoring and record keeping procedures, and corrective actions set forth in the HACCP plan. Validation also encompasses reviews of the records themselves, routinely generated by the HACCP system, in the context of other validation activities.

2. **Ongoing verification activities.** Ongoing verification activities include, but are not limited to:

- The calibration of process-monitoring instruments
- Direct observations of monitoring activities and corrective actions; and
- The review of records.

3. **Reassessment of the HACCP plan.** Every establishment should reassess the adequacy of the HACCP plan at least annually and whenever any changes occur that could affect the hazard analysis or alter the HACCP plan. Such changes may include, but are not limited to, changes in: raw materials or source of raw materials; product formulation; processing methods or systems; production volume; personnel; packaging: product distribution systems; or, the intended use or consumers of the finished product. One reassessment should be performed by an individual trained in HACCP principles. The HACCP plan should be modified immediately whenever a reassessment reveals that the plan no longer meets the requirements of the Food Code.

4. **Reassessment of the hazard analysis.** Any establishment that does not have a HACCP plan because a hazard analysis has revealed no food safety hazards that are reasonably likely to occur should reassess the adequacy of the hazard analysis whenever a change occurs that could reasonably affect whether a food safety hazard exists. Such changes may include, but are not limited to changes in: raw materials or source of raw materials; product formulation; processing methods or systems; production volume; packaging; finished product distribution systems or the intended use or consumers of the finished product.

Verification procedures
help make the HACCP plan
work correctly.

CHAPTER 4

The Peliminary Steps

Now that you have a general understanding of HACCP, let's get down to the specifics. Developing a HACCP plan starts with the collection of important information. This fact-finding process is called the Preliminary Steps.

1 Assemble the HACCP team.

2 Identify Products and Processes

3 Develop a complete list of ingredients, raw materials, equipment, recipes and formulations.

4 Develop a process flow diagram that completely describes your purpose.

5 Verify the process flow diagram.

In order to show you how an HACCP plan is put together, we are going to show you examples of filled-out HACCP development forms. The thought of filling out all these forms can be a bit overwhelming at first, however, it is a straightforward process.

Step 1:
Assemble the HACCP Team

YOUR FIRST TASK in developing a HACCP plan is to assemble your HACCP team. The HACCP team consists of individual(s) who will gather the necessary information for your HACCP plan.

The HACCP team needs to be aware of the following:

- Your product/process
- Any food safety programs you already have
- Food safety hazards of concern
- The seven principles of HACCP

In a very small facility, perhaps only one individual is available to be on the HACCP team. This is perfectly acceptable, however, you can get help from as many people as you need to make the team function effectively.

The HACCP team will begin by collecting scientific data. Remember, the team isn't limited to internal resources.

However you decide to approach it, your HACCP team is ultimately responsible for building your HACCP plan.

Working with the "HACCP Team" Form The Example Facility has six HACCP team members. One of whom is not only the general manager, but is also the owner. It is important to list all the team members and to state clearly what their HACCP team role is. (As you might think, filling out this form is realtively simple.)

[A note about the forms. As with all HACCP forms and logs, the person who is responsible for an activity (whether it be drafting the forms, or doing the monitoring) should be the one who signs and dates the form or log.]

The First Meeting

Who should be there, and what should we do?

Here's a sample agenda.

- First, describe your product -what it is and where it is going.
- Next, gather a complete list of ingredients

Step 1
HACCP Team Form

Team Members	Role
Cindy Jones	General Manager
Mary Weston	Quality Control
Mark Baker	Wet Room Supervisor
Susan Smith	Packing Supervisor
Joe Jones	Extension Service
Pam Smith	Local Microbilogist

Defveloped by: Barbara Jones Date: 21/02/21

Step 2:
Identify Products/Processes to be Covered

NEXT, make a complete listing of all the products and processes that must be covered under a HACCP plan. The foods should be categorized by the types of processes that must be covered. In addition, the requirements for reduced oxygen packaged foods limit the types of foods that can be packaged in this manner.

Product/Process Description Form
The following is an example of a format that could be used to list the products covered. This sample lists many types products and processes for this establishment - a typical store would not likely have all of these processes.

Product/Processes Covered

Store Name *General J's Market*

Street Address *123 XYZ Street*

City *Anytown* State *MN* Zip Code *55555*

Products/Processes Covered Under the HACCP Plan

Smoking/Curing

All Beef Summer Sausage, Ring Bologna, Smoked Turkey Drumsticks, Wieners, Snack Sticks, Beef Jerky, Bacon

Reduced Oxygen Packaging

All smokehouse products listed above

Sliced ham, sliced smoked turkey, sliced salami, hard cultured cheese (sliced and block), raw meats (cut and ground meat and poultry)

Food Additives

Acidified rice

Variances

Molluscan Shellstock sold from life support tanks

Sale of more than one tagged box of molluscan shellstock at any one time

Deviation of required cook times and temperatures for roast beef

Defveloped by: Barbara Jones Date: 21/02/21

Step 3:

Develop a Complete List of Ingredients, Materials, Equipment and Recipes/Fomulations

THE THIRD STEP is for the team to thoroughly review each product or process and write down all of the ingredients, materials and equipment used in the preparation or sale of a food and also to write down formulations or recipes that show methods and control measures that address the food safety concerns involved.

The ingredients list may be as simple as the recipe format listed below or may be more detailed as shown on the following page. As you can see on the following examples, ingredients and materials fall into several categories. If the category does not apply to your product/process, you don't have to write anything in that space.

[If you use pre-packaged or pre-blended ingredients such as a seasoning mix, you can list it by blend (mix) name and just staple that products information to the back of your Ingredients Form.]

Be sure a recipe is listed for every product you produce.

Ring Bologna
FULL BATCH
50 lbs pork trim
50 lbs beef trim
6 lbs (1 full package) of xyz brand bologna seasoning
4 oz (1 full package) of Quick Cure with sodium nitrite
10 lbs. of water
Casings - natural beef casing
Also list procedures for producing the product.

Smokehouse Operations Formulation/Recipe

Step 3
Ingredients and Raw Materials Form

Product/Process Name: *Fully cooked, Ready-to-eat*

Product/Examples: *Beef Jerky*

Meat/Poultry and Byproducts	Nonmeat Food Ingredients	Binders/Extenders
50 lbs. Beef Rounds		
Spices/Flavorings	**Restricted Ingredients**	**Preservatives/Addifiers**
__ oz. Garlic __ oz. Pepper (black) __ oz. Soy Sauce	__ oz. Sodium Nitrite	
Liquid	**Packaging Materials**	**Other**
__ lb. Tap Water	Vacuum Plastic Pouch Assorted Labels	

Defveloped by: Barbara Jones Date: 21/02/21

An additional requirement is to include a listing of all equipment and materials (such as packaging materials) used for each product produced or each type of process. This information can be written in list form and be categorized for the different processes.

Equipment List

Store Name *General J's Market*

Street Address *123 XYZ Street*

City *Anytown* State *MN* Zip Code *55555*

Smokehouse Operations Equipment List

- Walk-in Cooler: Brand ____________ Size ____________
  Other products/Operations Supported ____________
- Grinder: Brand ____________ Model ____________
- Mixer: Brand ____________ Model ____________
- Stuffer: Brand ____________ Model ____________
- Smokehouse: Brand ____________ Model ____________
  Smoke generator/liquid smoke ____________
- Digital Thermometer ____________
- Assorted measuring container, hand utensils, lugs, totes, etc. ____________

____

Reduced Oxygen Packaging Equipment List

- Slicer: Brand ____________ Model # ____________
- Vacuum Packaging Machine ____________
- Digital Thermometer ____________
- Assorted knives, tongs, trays, lugs, totes, hand utensils, etc. ____________

____

- Vacuum plastic pouch ____________
- Scale/labeling machine ____________

Step 4 & 5:
Develop and Verify a Process Flow Diagram

AT STEPS 4 AND 5 the team will create a document that will be used over and over again in the HACCP plan development process. The HACCP team needs to look closely at the production process and make a flow diagram that shows all the steps used to prepare the product. You don't need to include steps that are not directly under your control, such as distribution.
The flow diagram doesn't need to be complex. Looking at your facility's fl oor plan can help you visualize the process from receiving to shipping. To fi nd all the food safety hazards in your process you need to know exactly what steps that product/process goes through.
After the HACCP team has completed the flow diagram, it needs to be checked for accuracy. To do this, walk through the facility and make sure that the steps listed on the diagram realistically describe what occurs during the production process. If possible, have someone who didn't make the flow diagram do the "walk-through."

After the HACCP team completed their drawing, the flow diagram was checked, signed and dated. The form must be signed and dated again after it is checked/reviewed.

Steps 4 & 5
Process Flow Diagram Development & Verification Form

Product/Process Name *Beef Jerky/Heat Treated, Shelf Stable*

Flow Diagram:

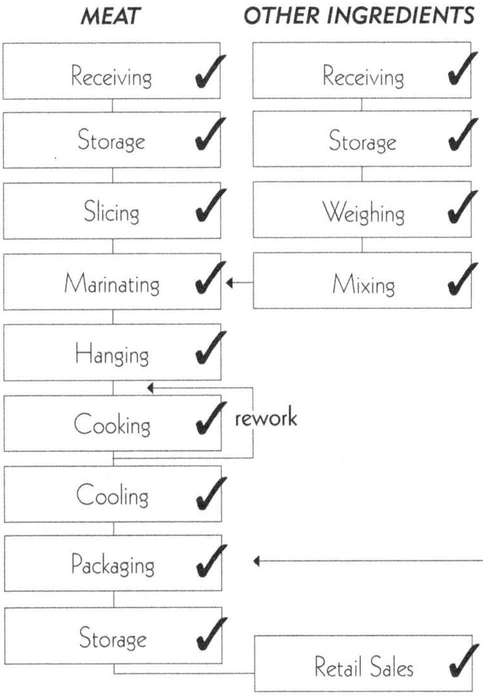

Defveloped by: Barbara Jones Date: 21/02/21
Verified by: Mary Weston Date: 23/02/21

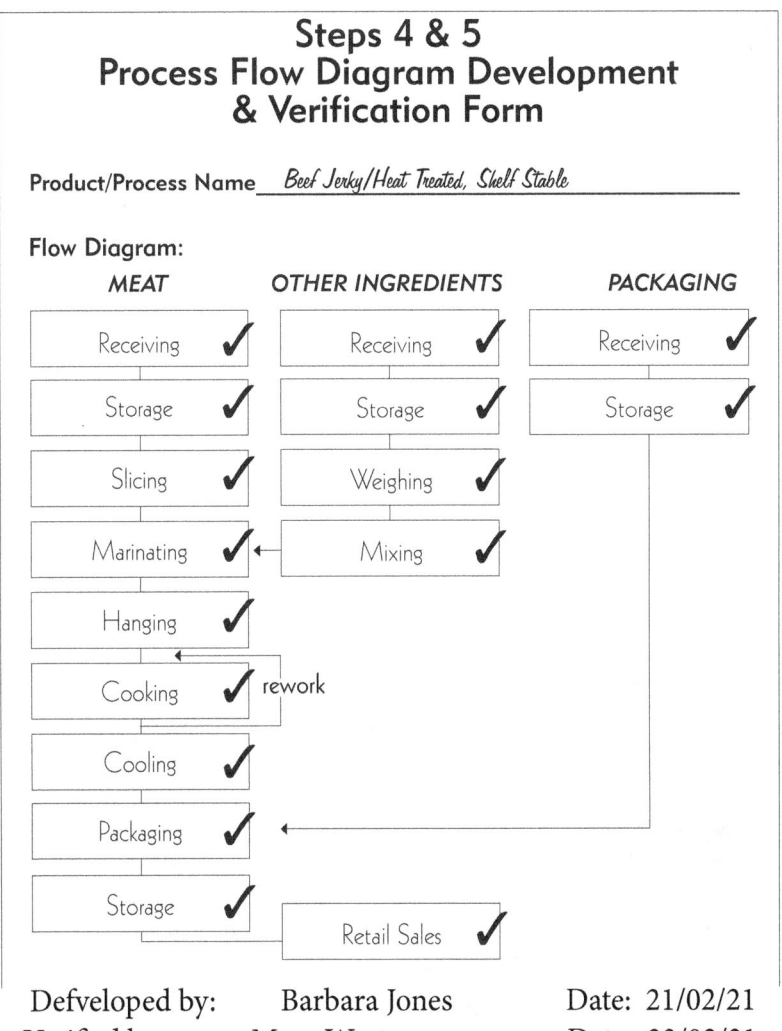

Conclusion:

Your work through the preliminary steps should have produced two tangible pieces of information:

1. A comprehensive list of ingredients and raw materials, and
2. A step-by-step production process breakdown, laid out simply in a flow diagram.

With this information you are now ready to proceed to the next stage: Utilizing the 7 Principles of HACCP.

CHAPTER 5

Utilizing the 7 Principles of HACCP

Understanding Hazards and Controls

This chapter is about using the seven principles of HACCP. Already you have gathered all of te specific information about our facilities paroducts and processes. Now you'll put that information to use. When you have worked throught the principles of HACCP, you'll have a complete HACCP plan.

Before we start with the first principal, we need to quickly review two important ideas; Food Safety Hazards and Preventative Measures. Hazards are defined as any biological, chemical or physical property that is reasonably lickly to cause food to be unsafe for human consumption.

Hazards are classified into these three categories: Biological, Chemical, and Physical.

Biological hazards can be bacteria, parasites, or viruses. Bacteria, parasites, or viruses that cause illness are called pathogens. In most cases, pathogens must grow or multiply in food to certain levels in order to cause foodborne illness. The following factors can affect the growth of pathogens:

Nutrients

Bacteria require food and water to carry on their life processes. Since what you are producing is a food product, nutrients are going to be available. Equipment that contains food residue can also be a nutrient source for bacteria.

Temperatures

Another essential factor that affects the growth of bacteria is temperature. Growth can occur over a wide range of temperatures from about 14°F to 194°F, but individual bacteria have much narrower temperature ranges for growth.

Time

It's not just the temperature that's the problem; it's the time at these temperatures that can affect growth of bacteria. The goal is to minimize the time of exposure of foods to temperatures where bacteria grow most quickly.

Moisture

The amount of available moisture in a food is measured as water activity. When substances like salt and sugar are added to water is tied up and is less available to the bacteria. The water activity of some foods is listed below:

Food	Water Activity
Fresh meats, fish, fruits, and vegetables	0.98 or above
Cured meat, processed cheese, bread	0.93 – 0.98
Dried meat, aged cheddar cheese	0.85 – 0.93
Cereal, flour, jam, nuts, salted fish	0.60 – 0.85
Chocolate, honey, noodles	0.60 or below

Most bacteria will not grow when the water activity is 0.85 or less. Many yeasts and molds can grow below this level but this is a spoilage concern and generally not a food safety concern.

Inhibitors

Foods can contain chemicals that are either natural or added that restrict or prevent growth of microorganisms. Salt is a good example of an added chemical that can inhibit growth of bacteria. Chemical preservatives like sodium nitrite, sodium benzoate, and calcium propionate can also inhibit the growth of microorganisms.

pH

pH shows how acid a food is. pH ranges from 0 – 14 with 7 being neutral. Foods with a pH of 4.6 and below are considered acid foods, like most fruit juices. Foods with a pH above 4.6 are said to be low acid, like meats and vegetables. Most bacteria don't grow very well in acid foods, so you can use pH to control the growth of bacteria. Generally, food is considered to be in a safe pH range when the fi nal pH is 4.6 or below.

Atmosphere

Some bacteria require a specific type of atmosphere for growth. Microorganisms are categorized as aerobes, anaerobes, facultative anaerobes and microaerophilic. Aerobes require oxygen and include such bacteria as Bacillus. Anaerobes grow only in the absence of molecular oxygen. These organisms include Clostridium. Facultative anaerobes can grow whether the environment has oxygen or not. Microaerophilic is a term applied to organisms, which grow only in reduced oxygen environments. Knowledge of the atmosphere surrounding the food is an especially important consideration in determining which pathogens are likely to be a problem.

Chemical Hazards

A wide variety of chemicals are routinely used in the production and processing of foods. Some examples of common types of chemicals are listed in the following table. While these types of chemicals may not be hazards if used properly, some can cause illness if not used properly. Therefore, the hazard analysis must consider whether any of these chemicals is used in a manner which creates a signifi cant food safety problem.

Physical Hazards

Physical hazards are represented by foreign objects or extraneous matter that are not normally found in food. The presence of these items typically result in personal injuries such as a broken tooth, cut mouth, or a case of choking. Examples of Physical hazards are found in the following Table. In some instances, physical contaminants may also include "filth" such as mold mats, insects, and rodent droppings. Although extraneous matter normally categorized as filth may not actually injure a consumer, some of these items can also contribute biological hazards. For example, rodents and their droppings are known to carry Salmonella species.

BACTERIA - CHARACTERISTICS OF GROWTH

Pathogens	Temperature for Growth (F°)	pH	Minimum Water Activity (A_w)
Bacillus cereus	39 – 131	4.3 – 9.3	0.92
Campylobacter jejuni	86 – 113.7	4.9 – 9.5	0.99
Clostridium botulinum	38 – 118	A: 4.6 E: 5.9	A: 0.94 E: 0.97
Clostridium perfringens	50 – 125	5.0 – 9.0	0.93
Escherichia coli	45 – 121	4.0 – 9.0	0.95
Listeria monocytogenes	31 – 113	4.4 – 9.4	0.92
Salmonella	41 – 115	3.7 – 9.5	0.94
Shigella	43 – 117	4.8 – 9.3	0.96
Staphylococcus aureus	45 – 122	4.0 – 10	0.83
Vibrios	41 – 111	4.8 – 11	0.94 – 0.97
Yersinis enterocolitica	30 – 108	4.2 – 10	0.95

EXAMPLES OF CHEMICAL HAZARDS	
Location	**Hazard**
Raw Materials	Pesticides, antibiotics, hormones, toxins, fertilizers, fungicides, heavy metals, PCB's
	Color additives, inks, indirect additives, packaging materials
Processing	Direct food additives - preservatives (high level of nitrates) flavor enhancers color additives
	Indirect food additives - boiler water additives peeling aids defoaming agents
Building and Equipment Maintenance	Lubricants, paints, coatings
Sanitation	Pesticides, cleaners, sanitizers
Storage and Shipping	All types of chemicals

EXAMPLES OF PHYSICAL HAZARDS	
Cause	**Source**
Glass	Bottles, jars, light fixtures, utensils, gauge covers, thermometers
Metal	Nuts, bolts, screws, steel wool, wire, meat hooks
Stones	Raw materials
Plastics	Packaging materials, raw materials
Bone	Raw materials, improper plant processing
Bullet/BB shot/Needles	Animals shot in field, hypodermic needles used for injections
Jewelry/Other	Rings, watches, pens, pencils, buttons, etc.

Preventative Measures are defined as: "Physical, chemical or other means that can be used to control an identified food safety hazard." The following tables provide examples of preventive measures for Biological, Chemical, and Physical Hazards.

EXAMPLES OF PREVENTATIVE MEASURES FOR BIOLOGICAL HAZARDS	
Pathogen	Preventive Measure or Control
Bacillus cereus	Proper handling and cooling temperatures of foods; thermal processing of shelf-stable canned food.
Campylobacter jejuni	Proper pasteurization or cooking; avoiding cross-contamination of utensils, equipment; freezing; atmospheric packaging.
Clostridium botulinum	Thermal processing of shelf-stable canned food; addition of nitrite and salt to cured processed meats; refrigeration of perishable vacuum packaged meats; acidification below pH 4.6; reduction of moisture below water activity of 0.93.
Clostridium perfringens	Proper handling and cooling temperatures of foods; proper cooking times and temperatures; adequate cooking and avoidance of cross-contamination by unsanitary equipment.
E-coli 0157:H7	Proper heat treatment ; prevention of cross contamination; proper refrigeration temperatures.
Listeria monocytogenes	Proper heat treatments; rigid environmental sanitation program; separation of raw and ready-to-eat production areas and product.
Salmonella spp.	Proper heat treatments; separation of raw and cooked product; proper employee hygiene; fermentation controls; decreased water activity; withdrawing feed from animals before slaughter; avoiding exterior of hide from contacting carcass during skinning; antimicrobial rinses scalding procedures; disinfecting knives.
Shigella	Proper heat treatment; proper holding temperatures; proper employee hygiene.
Staphylococcus aureus	Employee hygiene; proper fermentation and pH control; proper heat treatment and post-process product handling practices; reduced water activity.
Vibrios	Proper heat treatment; prevention of cross-contamination; proper refrigeration temperatures.
Yersinia enterocolitica	Proper refrigeration; heat treatments; control of salt and acidity; prevention of cross-contamination.

You should now be able to identify many types of hazards. You should also know where to begin looking for their preventative measures.

EXAMPLES OF PREVENTIVE MEASURES FOR CHEMICAL HAZARDS	
Hazard	Preventive Measure
Naturally-occuring Substances	Supplier warranty or guarantee; verification program to test each supplier's compliance with the warranty or guarantee.
Added Hazardous Chemicals	Detailed specifications for each raw material and ingredient; warranty or letter or guarantee fro the supplier; visiting suppliers; requirement that supplier operates with a HACCP plan.
In-Process Chemicals	Identify and list all direct and indirect food additives and color additives; check that each chemical is approved; check that each chemical is properly used; record the use of any restricted ingredients.

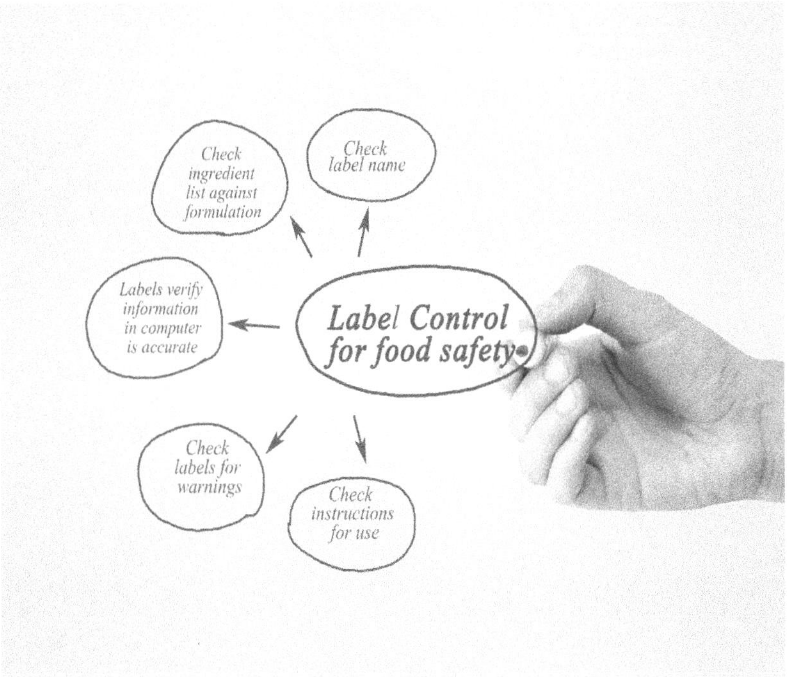

EXAMPLES OF PREVENTIVE MEASURES FOR PHYSICAL HAZARDS	
Hazard	Preventive Measure
Foreign objects in raw materials	Supplier's HACCP plan; use of specifications, letters of guarantee; vendor inspections and certification; in-line magnets; screens, traps, and filters; in-house inspections of raw materials.
Foreign objects in packaging materials, cleaning compounds, etc.	Supplier's HACCP plan; use of specifications, letters of guarantee; vendor inspections and certification, in-house inspections of raw materials.
Foreign objects introduced by processing operations or employee practices	In-line metal detectors; visual product examinations; proper maintenance of equipment; frequent equipment inspections.

Principle 1:

Hazard Analysis

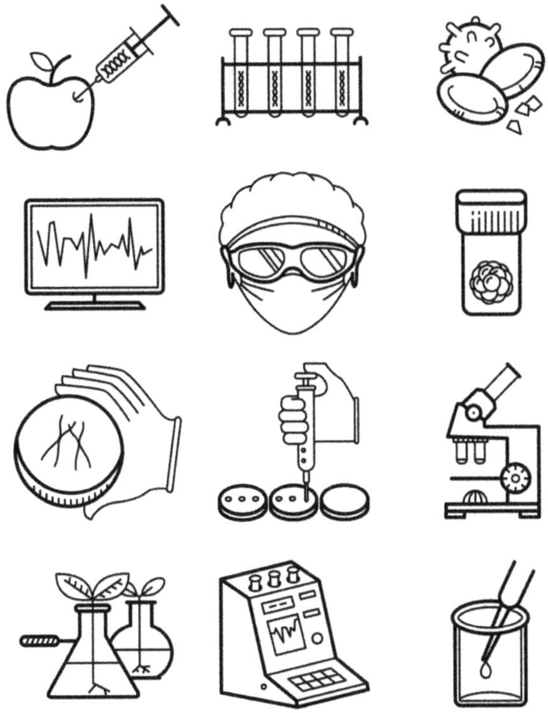

Preventive Measures:

When determining the appropriate preventative measure for an existing food safety hazard, keep in mind the wealth of regulatory, scientific, and historical support. Over the years, both industry and regulators have done a lot of work in identifying food safety hazards and preventative measures that can be used to control them in food production. Don't think that you have to go it alone in this search.

A thorough hazard analysis is one of the keys to building an effective HACCP plan. The hazard analysis process involves identifying hazards that are reasonably likely to occur in the absence of control and their preventive measures. In the first "Identifi cation" stage, the HACCP team identifies and lists food safety hazards that may be introduced or increased at each step in the production process.

Then, in the second "Evaluation" stage, each food safety hazards is evaluated based on how likely it is to occur. The term "reasonably likely to occur" is the ruler against which each hazard can be measured. Also during this evaluation stage the HACCP team investigates the appropriate preventative measures that will control the "likely to occur" food safety hazards.

[Hazards can vary greatly from one store to another due to differences in sources of ingredients, product formulations, processing equipment, processing methods, duration of the processes, and storage methods. Make sure that your hazard analysis takes into account what's unique about your establishment.]

Hazard Identification and Evaluation

The following steps can help you and the HACCP team get started conducting your hazard analysis.

1 Here are some questions you can ask yourself to better understand the hazard identifi cation process:

- Are additives or preservatives added to the product to kill or inhibit the growth of bacteria?
- Will the amount of acidic ingredients affect the growth/survival of bacteria?
- Does the product need to be refrigerated/frozen or kept dry in storage and during transit?

2 Second, look at the product ingredients that you listed earlier. In order to find all of the food safety hazards that are reasonably likely to occur, you need to know detailed characteristics about all the ingredients used in your process, as well as possible ingredient interactions.

Here are some questions you can ask about the ingredients:

- Could these ingredients contain any pathogenic bacteria, dangerous chemicals, or harmful physical objects?
- If contaminated or mishandled, could the ingredients or materials support the growth of pathogenic bacteria?
- Are hazardous chemicals used in growing, harvesting, processing or packaging an ingredient?
- Is this ingredient hazardous if used in excessive amounts?

3 Third, determine if any food safety hazards exist for each processing step listed in the process flow diagram.

Here are some questions you can ask for each production step:

- Could contaminants reach the product during this processing step?
- Could this step create a situation where an ingredient, work in process, or finished product becomes contaminated with pathogens?
- Could this step introduce a chemical or physical hazard into the product?

Possibilities for the three questions above include: worker handling, contaminated equipment or materials, crosscontamination from raw materials, leaking valves or pipes, splashing, etc.

- Could bacteria multiply during this process step to the point where they became a hazard? Consider product temperature, hold temperature, etc.

KEEP GOOD NOTES

A summary of the HACCP team meetings and the reasons for each decision during the hazard analysis should be kept for future reference. These documents will be a great help to you when you have to review and update your hazard analysis and HACCP plan.

Finding Preventive Measures

Now that you have a good idea of what you're looking for in the way of hazards. Use the example tables of preventive measures on pages to use as a reference to fi nd out some ways to keep those hazards under control.

It is sometimes the case that more than one preventive measure may be required to control a specifi c hazard, or that more than one hazard may be controlled by one preventive measure. As you go through the hazard analysis, you may recognize preventive measures already in place in your production processes.

The key to a successful hazard analysis is to link the preventive measures to the food safety hazards you have just identified.

Working with the "Hazard Analysis" Form

The form is structured so that the three food safety hazard categories (chemical, biological, physical) are addressed in each of the four questions. Don't forget that you need to fill out the top of the form with the appropriate information, such as the product/process name, and the process steps from the flow diagram. You also need to sign or initial and date the form when it's complete.

The first production step we're going to look at is receiving meat.

1 For the first question all you need to do is state what food safety hazards are present at that step.

2 The second question asks you to decide whether or not the hazard is reasonably likely to occur at that step.

3 The third question is where you explain why you answered "Yes" or "No", to the question of "reasonably likely to occur." For the biological hazard they assume that the bacteria is on the meat prior to arrival, so that it continues to be a potential hazard.

[This "historical" basis for deciding whether a food safety hazard is "reasonably likely to occur" is perfectly legitimate. If your facility has a clean track record regarding a particular hazard, it's fine to include that information in your HACCP plan. All information must be documented.]

4 The final question on the hazard analysis form is the place where you write the specific preventive measure(s) that will control the hazard you said was likely to occur. With each shipment of meat the Example Facility receives they feel that the "Letter of Guarantee" from their supplier reasonably assures them the meat has been kept at a temperature adequate to control bacterial growth. However, just because they have one preventive measure hasn't stopped them from also having a second preventive measure. They also visually check the condition and temperature of the truck and meat products, to make sure everything meets their standards.

HACCP Principle 1
Hazard Analysis Form

Product/Process Name: *Beef Jerky/Heat Treated, Shelf Stable*

Process Step from Flow Diagram: *Receiving Meat*

C: CHEMICAL	B: BIOLOGICAL	P: PHYSICAL
List the Hazards:		
Pesticides	Pathogens	Plastic
Hormones		Bone Fragments

Is the hazard reasonably likely to occur?

C: CHEMICAL	B: BIOLOGICAL	P: PHYSICAL
❑ Yes ☒ No	☒ Yes ❑ No	❑ Yes ☒ No

What is the basis for your decision?

C: CHEMICAL	B: BIOLOGICAL	P: PHYSICAL
No evidence of any historical occurence at this facility.	Loss of control in time/temp can promote harmful bacteria growth.	No evidence of any historical occurence at this facility fromthis product/source.

What preventative measures can be applied at this step to prevent, eliminate or reduce the hazard to an acceptable level?

Collect "Letter of Guarantee" from supplier that stipulates your requirements. If exceeds limits, product won't be accepted from supplier.

Defveloped by: Barbara Jones Date: 21/02/21

The second production step we're going to look at is cooking.

1 List the hazards. The Example Facility listed a chemical hazard of sanitizing chemicals because it's possible that traces of these substances could be on the equipment from the last time it was cleaned. They also listed a biological hazard because bacteria is unavoidable on all raw meat.

[If you don't find a particular type of hazard at a step it's okay to write "Non Identified" as the Example Facility did.]

2 Is it "reasonably likely to occur"? They answered "No" for the chemical hazard, and "Yes" for the biological hazard.

3 What is the basis for your decision? The Example Facility decided the sanitizing chemicals wouldn't be a hazard likely to occur because their proper use is thoroughly covered by existing Sanitation Standard Operating Procedures (SSOP'S). They decided "Yes" for the biological hazard for the same reason as in the preceding process step.

[When working on your HACCP plan, you might want to revisit your SSOP's]

4 What are the preventive measures? The Example Facility identified two preventive measures, cooking and water activity reduction for the biological hazard. They said this is because the cooking and the water activity reduction will help to reduce the hazard.

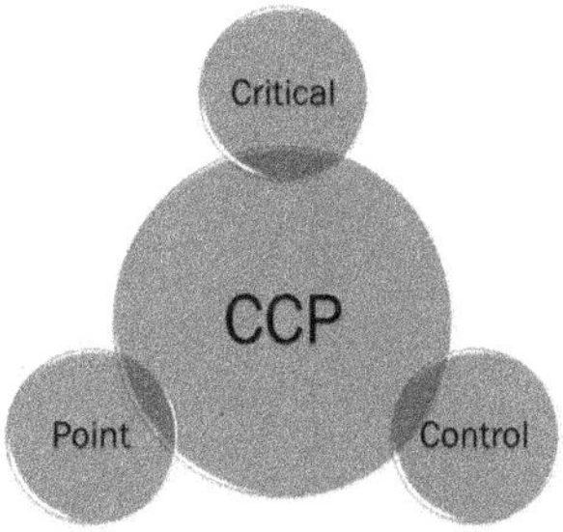

HACCP Principle 1
Hazard Analysis Form

Product/Process Name: Beef Jerky/Heat Treated, Shelf Stable

Process Step from Flow Diagram: Cooking

C: CHEMICAL	B: BIOLOGICAL	P: PHYSICAL
List the Hazards:		
Residues of sanitizing chemicals	Pathogen survival and growth in finished product.	(None Identified)

Is the hazard reasonably likely to occur?

C: CHEMICAL	B: BIOLOGICAL	P: PHYSICAL
☐ Yes ☒ No	☒ Yes ☐ No	☐ Yes ☐ No (None Identified)

What is the basis for your decision?

C: CHEMICAL	B: BIOLOGICAL	P: PHYSICAL
Proper use will address this issue.	Loss of control in time/temp or moisture level can promote harmful bacteria growth.	(None Identified)

What preventative measures can be applied at this step to prevent, eliminate or reduce the hazard to an acceptable level?

Smokehouse temperature is 190°F.

Defveloped by: Barbara Jones Date: 21/02/21

The third production step we're going to look at is cooling.

1 List the hazards. The Example Facility listed the biological hazard of cross-contamination because any time when you have raw and finished product in the same facility the possibility for the raw product to cross-contaminate the fi nished product exists. The Example Facility also listed plastic as a physical hazard because this is the step where they "Pull" the jerky strips off the cooking trees into large plastic barrels.

2 Is it "reasonably likely to occur"? The Example Facility answered, "No" for the biological, and "No" for the physical.

3 What is the basis for your decision? The Example Facility said that the biological hazard was not likely to occur because the raw and cooked products are strictly kept apart as called for in their SSOP's. They said "No" to the physical hazard because the plastic barrels that are used are made of an extremely sturdy type of plastic and there's never historically been a problem with plastic shavings at this facility getting into the jerky.

4 What are the preventive measures? There aren't any preventive measures listed here because no food safety hazards were found to be reasonably likely to occur.

HACCP Principle 1
Hazard Analysis Form

Product/Process Name: *Beef Jerky/Heat Treated, Shelf Stable*

Process Step from Flow Diagram: *Cooling*

	C: CHEMICAL	B: BIOLOGICAL	P: PHYSICAL
List the Hazards:	(None Identified)	Pathogen cross-contamination	Plastic
Is the hazard reasonably likely to occur?	☐ Yes ☐ No (None Identified)	☐ Yes ☒ No	☐ Yes ☒ No
What is the basis for your decision?	(None Identified)	SSOP's for separtation	No evidence of any historical occurence at this facility.

What preventative measures can be applied at this step to prevent, eliminate or reduce the hazard to an acceptable level?

Defveloped by: Barbara Jones Date: 21/02/21

Principle 2:

Identify Critical Control Points

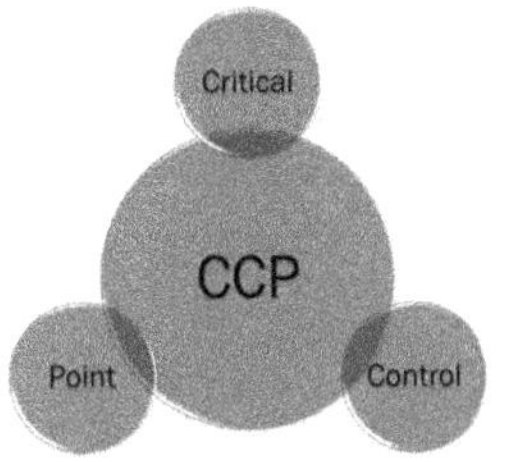

Numbering your CCP's:
Once you've been through your entire production process and have successfully identified all the CCP's there's one more thing you need to do to get that CCP set up. You need to organize them. Feel free to do this anyway that works for your business.

One easy way to accomplish this is to develop a simple numbering system. It's a good idea to always write "CCP" before the numbers - this can make your documents easier to understand. For instance you could write it like: CCP#1, CCP#2, CCP#3.

Also remember that you could have more than one CCP (for a designated food safety hazard) at a given process step or one CCP may control more than one hazard. In this case you might want to include the letter "B", "C" or "P" to identify whether it is a biological, chemical or physical hazard. For example: CCP#1B, CCP#1C, CCP#2P, CCP#2C.

A critical control point is defined as "A point, step or procedure in a food process at which control can be applied and is essential to prevent or eliminate a food hazard or reduce it to an acceptable level." Everything in your HACCP plan revolves around the proper identifi cation of CCPs. Some of the most common CCPs are:

• Chilling or freezing to a specifi ed temperature to prevent bacteria from growing.
• Cooking that must occur for a specific time and temperature in order to destroy bacteria.
• Prevention of cross-contamination between raw and cooked product.
• Certain processing procedures, such as filling and sealing cans, mixing and spicing, etc.
• "pH".
• Holding at proper refrigeration temperatures.

These are just a few examples of possible CCPs. Different facilities, preparing the same food, can identify different food safety hazards and different critical control points. Usually no two stores have the same floor plan, equipment, or ingredients. The CCPs you identify will reflect the uniqueness of your processing facility.

One of the tools used to help determine critical control points is a "CCP Decision Tree." The use of a Decision Tree to identify signifi cant hazards is not necessary for you to meet regulatory requirements. However, the thought process may be helpful for your team; you want to make sure that your HACCP system meets regulatory requirements.

The Example Facility used the CCP Decision Tree to take a closer look at both of the steps in their process where they determined food safety hazards were reasonably likely to occur. [Go ahead and read the four questions on the form and then we'll look at each one in detail. Again, this approach is not necessary to meet regulatory requirements.

The first step they looked at was receiving meat.

Question 1a

The Example Facility answered "Yes" because the "Letter of Guarantee" from the supplier, and checking the temperature of the truck and products are the preventive measures for this biological hazard.

Question 1b

If you answered "Yes" for question 1a, then you don't need to worry about question 1b. (If you haven't yet identified a preventive measure for a food safety hazard, question 1b will not let you move down the CCP Decision Tree until you do.)

Question 2

This question asks whether or not this step "prevents, eliminate, or reduces" to acceptable levels, the food safety hazard you are working with. The Example Facility said "No" because simply receiving the meat doesn't mean the hazard is controlled.

Question 3

The Example Facility said "Yes" here because, if not controlled, the biological hazard could get worse.

Question 4

Here the HACCP team must decide if this step is the last point at which control could be applied to the hazard. In this case the Example Facility found that, in fact, a later step (i.e. cooking) could control this biological food safety hazard. This process step was not a CCP.

Critical Control Point Decision Tree

For the production of cooked products. Process Step Receiving Meat

Question 1A
Do preventative measures exist for the identified hazards?
If no - go to Question 1B.
If yes - go to Question 2. Yes, go to Question #2

Question 1B
~~Is control at this step necessary for safety?~~
~~If no - not a CCP.~~
~~If yes - modify step, process or product~~
~~and return to Question 1.~~

Question 2
Does this step eliminate or reduce the likely occurence of a hazard(s) to an acceptable level?
If no - go to Question 3.
If yes - CCP. No

Question 3
Could contamination with identified hazard(s) occur in excess of acceptable levels or could these increase to unacceptable levels?
If no - not a CCP.
If yes - go to Question 4. Yes

Question 4
Will a subsequent step eliminate the identified hazards or reduce the likely occurence to an acceptable level?
If no - CCP.
If yes - not a CCP.

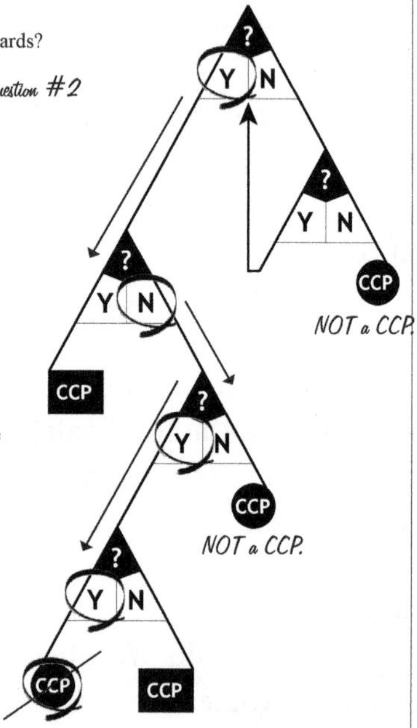

Results:
Yes - so it's NOT a CCP.

BIOLOGICAL	CHEMICAL	PHYSICAL
❑ CCP#________ ☒ Not a CCP	❑ CCP#________ ❑ Not a CCP	❑ CCP#________ ❑ Not a CCP

Defveloped by: Barbara Jones Date: 21/02/21
Verified by: Mary Weston Date: 23/02/21

The second step they looked at was cooking.

Question 1a
The Example Facility answered "Yes" here because they had identified the preventive measure of cooking (i.e. time and temperature) for this step.

Question 1b
As in the receiving example, move on to question 2.

Question 2
The Example Facility said that "Yes" cooking would eliminate the hazard at this step. They stopped here at question 2 because they reached a positive result...their CCP. Thus, there wasn't any need to go on to

questions 3 and 4.
[After finding all the CCP's in your process, the HACCP team needs to organize them. At the bottom of the CCP Decision Tree Form the Example Facility named the cooking CCP "CCP#01B". The "01" tells them what number the CCP is, and the "B" tells them it is a biological food safety hazard.]

Critical Control Point Decision Tree

For the production of cooked products. Process Step Cooking

–

Question 1A

Do preventative measures exist for the identified hazards?

If no - go to Question 1B.

If yes - go to Question 2. Yes, go to Question #2

~~*Question 1B*~~

~~Is control at this step necessary for safety?~~

~~If no - not a CCP.~~

~~If yes - modify step, process or product and return to Question 1.~~

Question 2

Does this step eliminate or reduce the likely occurence of a hazard(s) to an acceptable level?

If no - go to Question 3. Yes,

If yes - CCP. Identified as a CCP

Question 3

Could contamination with identified hazard(s) occur in excess of acceptable levels or could these increase to unacceptable levels?

If no - not a CCP.

If yes - go to Question 4.

Question 4

Will a subsequent step eliminate the identified hazards or reduce the likely occurence to an acceptable level?

If no - CCP.

If yes - not a CCP.

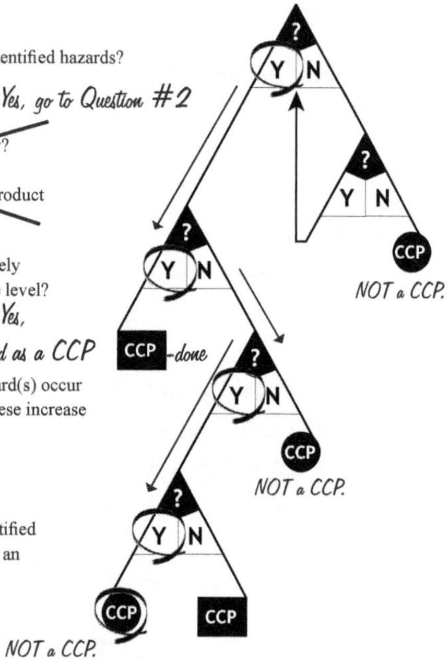

Results:

BIOLOGICAL	CHEMICAL	PHYSICAL
☒ CCP# #01B ❑ Not a CCP	❑ CCP# ❑ Not a CCP	❑ CCP# ❑ Not a CCP

Defveloped by: Barbara Jones Date: 21/02/21

Verified by: Mary Weston Date: 23/02/21

Principle 3:

Establish Critical Limits for Each Critical Control Point

Here are some controls commonly used as preventative measures.

• Time and Temp - The temperature "danger zone" for biological hazards is between 40°F and 140°F. Bacteria grows fast! They have the ability to multiply rapidly. Knowing this shows that controlling how long the product is in the danger zone (if at all) presents itself as an extremely effective critical limit.

•pH - The pH of a food product is the level of its acidity or alkalinity. The pH is measured on a scale of 0 to 14 . The middle of the scale, pH=7.0, is considered neutral. Altering a food product's pH, such as adding an acidic substance like vinegar or soy sauce will decrease the growth rate of the bacteria.

• Water Activity - In addition to warm temperatures and a median pH, bacteria also need water to grow. Water activity (AW) refers to the amount of water in a food product that is available, or free, for bacteria to use for growth and multiplication. Solutes (salts and vinegars), as well as dehydration, decrease the available water and can reduce bacterial growth.

A critical limit is defined as "The maximum or minimum value to which a physical, biological, or chemical hazard must be controlled at a critical control point to prevent, eliminate, or reduce to an acceptable level the occurrence of the identified food safety hazard." You can think of a critical limit as a boundary of safety for a CCP. The critical limit is the numerical value that must be reached to assure that hazards have been controlled. An example would be that "all sausage products must be cooked to 155 ° F for 15 seconds."

Each CCP will have at least one (possibly more) preventive measures that need to be controlled to assure this prevention, elimination or reduction of food safety hazards. To be effective, each critical limit should be:

1. Based on proven factual information.

A few ways that information and recommendations for appropriate limits can be obtained are: from regulatory requirements, scientific literature, and consultation with experts. If regulatory requirements exist they must be met or exceeded.

2. Objectives are measurable or observable, such as time and temperature.

3. Appropriate and reasonable for the food product and operation.

You should consider the type of equipment, the volume of product being produced, how the critical limit will be monitored and frequency of monitoring.

4. Specifics

When drafting your critical limits be specific in your language. Use action words, and be specific when naming people and equipment. An example could be "bake, uncovered in preheated 350°F oven to an internal temperature of 165 ° F for 15 seconds."

The HACCP team will fi nd that many critical limits for your identifi ed CCP's have already been established.

In some cases you'll need more than one critical limit to control a particular hazard. For example, the typical critical limits for cooked beef pattied are time/temperature, patty thickness, and conveyor speed. It is important that you identify all the critical limits for each of your products.

Making sure each Critical Control Point has critical limits is the responsibility of each establishment. The HACCP team may want to get help from outside HACCP experts when establishing critical limits. Remember that the critical limits must be able to maintain control over the food safety hazard. Once the team has identified all the limits, enter them onto the Critical Limits form.

Working with the "Critical Limits" Form

For each CCP the Example Facility has a separate page of critical limits.

1. Under the "Limit" heading.
The Example Facility noted an internal temperature of 165°F for 15 seconds as the established critical limit. They then decided that the preventive measure of cooking at 190° F oven temperature for 3 hours would satisfy the critical limit.

2. Under the "Source" Heading.
The Example Facility's first source is regulatory and scientific. They decided to take the established regulatory limits and use them, but then they also sent out samples of their fi nished product to be scientifically analyzed. The results of the lab tests confirmed that their critical limits were enough.

[The source is the "evidence" that backs up your critical limits. The source provides that the critical limits you cite will effectively control the food safety hazards. Sources for critical limits can be scientific, regulatory or historical. The HACCP team has to fi nd at least one source for each of your critical limits, but you can always put more if you want.]
When determining your critical limits make sure you file your supporting documentation with your HACCP plan. This documentation will help validate that the limits have been properly established. These could be things such as letters from outside HACCP experts, or scientific reports, or lab test results. By holding onto these supporting documents you also provide verification material when needed.

HACCP Principle 3
Critical Limits Form

Product/Process Name: Beef Jerky/Heat Treated, Shelf Stable

Process Step/CCP: Cooking CCP#01B

Critical Limits
Limit - (Time, Temp, pH, etc.)

Internal temperature: 165 degrees Fahrenheit for 15 seconds.

Preventive Measure: Oven temperature: 190 degrees Fahrenheit for 3 hours.

Source - (cite a regulation, scientific document, other resource)

Meets regulatory requirements

Laboratory tests and results

Defveloped by: Barbara Jones Date: 21/02/21

Principle 4:

Establish Monitoring Procedures

Remembering Your Monitoring

The key to effective and reliable monitoring is to keep it simple and build it into the employees' normal routines. When establishing a time for the actual monitoring procedure, allow some flexibility. For example, if you say you will monitor a CCP at 10 a.m. and the person is not there at exactly 10 a.m., you could be opening yourself up for problems. It is suggested that you specify a period of time during which monitoring will occur. For example, write your time as "10a.m. +/-- 10 minutes"or "between the time period of 10 a.m. and 10:15 a.m."

Monitoring involves a series of observations and/or measurements that are used to make sure a CCP is under control. The HACCP team can think of monitoring activities as the checks-and-balances for each CCP. When someone monitors, they are "checking to see" that the critical limits are being met.

What are the 3 things monitoring can do for you?

• shows you when a deviation from a critical limit has happened. For example, an employee tests the temperature of some beef patties and discovers that the internal temperature has gone above the established critical limit of 40°F. If not caught here, this would be a potentially serious health risk to consumers.

• helps you identify trends in your process that will allow you to predict a loss of control at a CCP. For example, a facility may monitor the temperature of a cold storage area at 6 am., 8 a.m., and 10 a.m. Each time, the temperature is within acceptable limits, but it is steadily climbing toward the high end of the range. This information points towards a trend, and the facility should take action to prevent the temperature from exceeding the critical limits.

• produces written records for use in future HACCP plan verification steps. Written monitoring records will prove very valuable to your operation, should a serious problem along the production line occur. The records you keep prove that your company has established and carried out effective monitoring techniques.

Monitoring procedures can be thought of as continuous or non-continuous.

• Continuous monitoring is the constant monitoring of a critical control point.
• Non-continuous monitoring is the scheduled monitoring of a critical control point.

Continuous monitoring is always preferred when feasible. Continuous monitoring at a CCP is usually done with built-in measuring equipment, such as a recording thermometer used at a cooking step. This type of monitoring is preferred because it yields a permanent record. To make sure these activities stay accurate, you need to regularly check the monitoring equipment to make sure that it is calibrated correctly.

If continuous monitoring isn't feasible for your CCP then the HACCP team will need to establish noncontinuous monitoring procedures. Non-continuous doesn't mean random. The team should decide in the development phase what the monitoring schedule should be. When you use non-continuous monitoring, make sure that it's scheduled often enough to keep the food safety hazards under control. Expert advice from people with knowledge of practical statistics and statistical process control will be important in making your decisions. Types of non-continuous monitoring procedures include visual examinations, monitoring ingredient specifi cations, measurements of pH or water activity (Aw), taking product temperatures, etc.

Who's Responsible?

Make sure to assign a specific person to be responsible for the monitoring of a CCP. The Example Facility has a designated shift leader/cook who is responsible for monitoring the cooking CCP. The person who actually does the monitoring must be the person who signs and dates all the records at the time of monitoring.

Monitoring will be most effective when:

• The HACCP plan clearly identifies the employee(s) responsible for monitoring.
• Employees are trained in the proper testing procedures, the established critical limits, the methods of recording monitoring results, and the actions to be taken when critical limits are exceeded.
• Employee(s) understand the purpose and importance of monitoring.

The last step in establishing your monitoring procedures is to develop the Monitoring Log(s) where the monitoring person will record the date for each CCP. Due to the variety of monitoring procedures, the HACCP team may need to developed different logs to record the monitoring data at different CCP's. When your HACCP system is up and running, you will use these logs to track the day-to-day HACCP activities.

Working with the "Monitoring Procedures" Form

The form that is shown as an example on the next page is to be used as a tool in the development of your HACCP plan. The information on this form is the "Who, What, When and How" of monitoring.

For the Example Facility:

- The Who is the cook on duty.
- The What is the temperature of the oven.
- The When is non-continuously - every 60 minutes, (+ 5 minutes), and
- The How is with the oven temperature gauge.

The Example Store felt this type of non-continuous monitoring would be effective because of the consistent heat environment of the oven. Their logic was that if the temperature taken at the beginning and end of the cooking cycle was the same, it could reasonably be assumed that it was okay for the whole cooking cycle.

HACCP Principle 4
Monitoring Procedures Form

Product/Process Name: *Beef Jerky/Heat Treated, Shelf Stable*

Process Step/CCP: *Cooking* *CCP #01B*

Monitoring Procedures - (Who, What, When, How)

The cook on duty records the oven temperature at intervals of 60 minutes, (± 5 minutes) starting when a "lot" is placed in the oven and ending when the "lot" is removed from oven. Each oven is monitored individually using an oven temperature gauge.

Defveloped by: Barbara Jones Date: 21/02/21

Principle 5:

Establish Corrective Actions

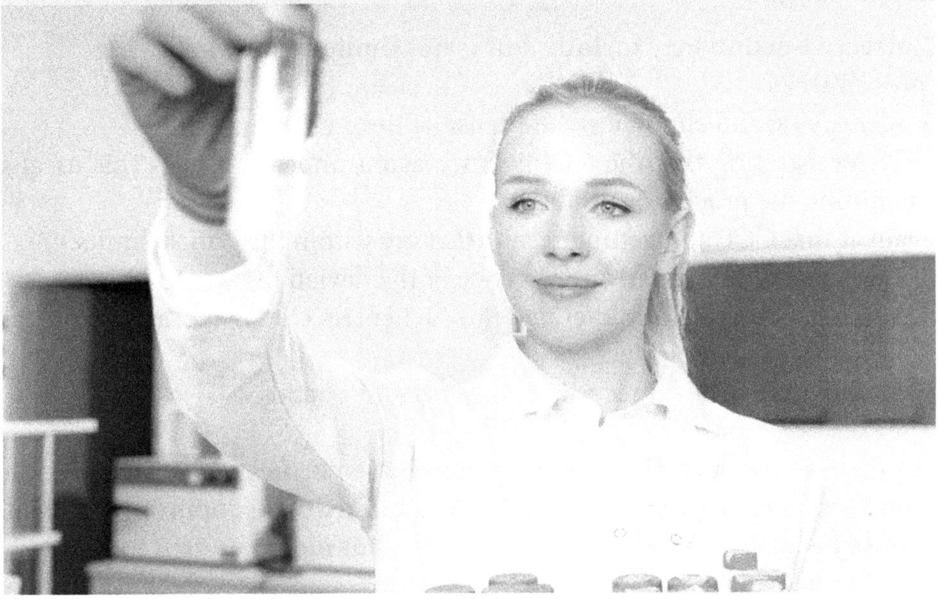

Stopping Production

The more ownership the employees feel they have in the HACCP system, the more effective they will be in ensuring that your facility produces safe food.

One idea is to empower the person responsible for monitoring to be able to stop production when and if a deviation occurs. This accomplishes two important functions.

• First, it prevents the potentially hazardous product from continuing down the production line.

• Second, it makes timely communication easier; thus you find out what's happening in your facility as soon as possible.

Corrective Action can be defined as "Procedures to be followed when a deviation occurs." A deviation is defined as a "failure to meet a critical limit."

Deviations can and do occur. After the HACCP team has established strict monitoring procedures, the next step is to draft corrective actions to be taken immediately when there is a loss of control at a CCP.

Corrective action may include, but is not limited to the following procedures:

1. Identifying and eliminating the cause of the deviation,
2. Demonstrating that the CCP is once again under control. (This means examining the process or product
again at that CCP and getting results that are within the critical limits.),
3. Taking steps to prevent a recurrence of the deviation,
4. Making sure that no adulterated product enters commerce, and
5. When to discard product.
6. Maintaining detailed records of the corrective actions.

If a deviation occurs that is not covered by a specific corrective action in your HACCP plan, or if some unforeseen hazard arises, appropriate steps should be taken. These steps shall include, but not be limited to:

1. Segregate and hold any affected product until its acceptability can be determined.
2. Determine the acceptability of the affected product for distribution.
3. Do not allow product that is injurious to health or is otherwise adulterated to enter commerce.
4. Reassess and, if necessary, modify your HACCP plan to properly address this type of deviation in the future.
5. Maintain detailed records of your actions.

Some examples of corrective actions are:

- Changing the process and holding the product for further evaluation.
- Empowering the monitoring personnel to stop the line when a deviation occurs. They should have the authority to hold all "lots" of a product not in compliance.
- Rely on an approved alternate process that can be substituted for one that

is out of control at the specific CCP.
• Additional cooking time.
• Quickly cooling product.

Whatever type of corrective actions the HACCP team establishes, records for each one need to be kept that include:

• That the deviation was identified.
• The reason for holding the product, the time and date of the hold, the amount of the product involved, and the disposition and/or release of the product.
• The actions that were taken to prevent the deviation from recurring.
• The dated signature of the employee who was responsible for taking the corrective action.

As with monitoring logs, the HACCP team also needs to develop the log(s) for the corrective action results.

Working with the "Corrective Action Procedures" Form

The Example Facility's corrective action form outlines exactly what they think should be done if a problem occurs with the CCP#01B.

• **Under the "Problem" heading.**
They state the critical limit that has been established for this CCP.

• **Under the "Disposition of Product" heading.**
If a deviation occurs, they have noted that the initial disposition would be to hold the product "lot", and try to rework it if possible. The "rework" would consist of fi xing the temperature and re-cooking the jerky.

• **Under "Corrective Action Procedures/Steps" heading.**
As you can see, the Example Facility listed quite specific corrective actions for this CCP. Their directions are written concisely, and in the order they should be performed.

• **Under the "Who is Responsible" heading.**
They are specific in naming a particular person.

• Under the "Compliance Procedures" heading.

The Example Facility has projected that if this deviation happens at this CCP it will probably be because something went wrong with the thermostat in the oven. They list here what will probably need to be done to make sure this doesn't happen again. (If this deviation were to actually happen, the monitoring person would write on the corrective action log what he or she did to fix the problem, and what they did to make sure it wouldn't happen again.)

HACCP Principle 5
Corrective Action Procedures Form

Product/Process Name: Beef Jerky/Heat Treated, Shelf Stable

Process Step/CCP: Cooking CCP # 01B

Problem - (Critical limit exceeded)
Oven temp. below 190 degrees Fahrenheit

Disposition of product - (Hold, Rework, Condemn)
Hold, rework if possible.

Corrective action procedures/steps 1. Identify and segregate affected product, place on hold.
2. Rework if possible, otherwise condemn product: Reestablish correct cooking procedures (i.e. fix oven temp. settings, or move product to other oven for rework.)
3. Determine cause of deviation: broken oven thermostat.
4. Take steps to prevent recurrence: recalibrate/replace thermostat
5. Notify Quality Control Supervisor a.s.a.p.

Who is responsible for performing these corrective actions? John Smith - Cook on duty

Compliance procedures
Recalibrate/Replace oven thermostat.
Monitor CCP as usual during rework.

Defveloped by: Barbara Jones Date: 21/02/21

Principle 6:

Establish Record Keeping Procedures

Tips on Designing Records
One way to approach development of the recordkeeping requirements of your HACCP system is to review the records you already keep, and see if they are suitable, in their present form or with minor modifications, to serve the purposes of your HACCP system. The best recordkeeping system is usually the simplest one that can easily be integrated into the existing operation.

The records you keep for HACCP can make all the difference! Good HACCP records - meaning that they are accurate and complete - can be a great help to you. Here's why:

• Records make it possible to trace ingredients, in-process operations, or finished products, should a problem occur.
• Records help you identify trends in your production line.
• Records serve as written documentation of your facility's compliance with the HACCP regulations.

Well maintained records protect both your customers and YOU.

Your HACCP records should include your development forms and your daily logs for each CCP. You should also keep your hazard analysis development forms, your CCP determination sheets, a list of critical limits for each food safety hazard, clear corrective action instructions, and a copy of your compiled HACCP plan. When first establishing your recordkeeping procedures, it's better to think of the different kinds of records you'll need in two ways.

First, there are records that are used for development for archival purposes; such as your Hazard Analysis, and your CCP decision making tool.

Second, there are records that you will work with on a day-to-day basis. These are the logs we've been discussing such as the monitoring or corrective action logs. As we've said before, the HACCP team will need to create these logs for each CCP in your process.

Regardless of the type of record, all HACCP records must contain at least the following information:

• Title and date of record.
• Product identifi cation,
• Signature of employee making entry,
• A place for the reviewer's signature, and
• An orderly manner for entering the required data.

Working with the "Recordkeeping Procedures" Form

- **Under the "Records" heading.**

You can see that the Example Facility has filled out their Recordkeeping Form making sure to list both the development forms (the hazard analysis), and the logs.

[One last note about the records you keep. When developing and working with your forms and logs remember to use ink (ballpoint pen) - no pencils. On all records, whenever you make a change, mark through the original and initial. Do not erase, white out, or mark the original so that it is unreadable.]

HACCP Principle 6
Recordkeeping Procedures Form

Product/Process Name: *Beef Jerky/Heat Treated, Shelf Stable*

Process Step/CCP: *Cooking* *CCP # 01B*

Records

Name and Location		
Name: Hazard Analysis Location: Office File Cabinet	Name: HACCP Plan Review Sheet - For each CCP Location: Oven Room Wall	Name: Monitoring Log - For each CCP Location: Oven Room Wall
Name: Deviation / Corrective Action Log Location: Oven Room Wall	Name: Process - Monitoring Equipment Calibration Log - For each CCP Location: Oven Room Wall	Name: Verification Procedures & Results Log - For each CCP Location: Oven Room Wall

Defveloped by: Barbara Jones Date: 21/02/21

Principle 7:

Establish Verification Procedures

Your team needs to decide on what procedures the facility will perform to verify that the HACCP system is working effectively and how often these actions will be performed. Verification uses methods, procedures, or tests in addition to those used in monitoring to see whether the HACCP system is in compliance with the HACCP plan or whether the HACCP plan needs modifi cation. There are three types of verification. These are initial validation, ongoing verifi cation, and reassessment of the HACCP plan.

Initial Validation

Validation is defined as" the specific and technical process for determining that the CCP's and associated critical limits are adequate and sufficient to control likely hazards." The initial validation of your HACCP plan is the process by which your establishment proves that what is written in the HACCP plan will be effective in preventing, eliminating, or reducing food safety hazards. This validation activity is the exclusive responsibility of your establishment.

You carry out this validation by gathering evidence that supports your HACCP plan. The data you bring together can come from many sources. Such sources may include scientific literature, product testing results, regulatory requirements, and /or industry standards. Companies have a lot of flexibility in the compilation of this information in regards to the sources and the amounts of such data.

[Most likely, you already have the majority of the validation information you need. When you conducted your hazard analysis and researched the sources for your critical limits, you were collecting data that could also be used to validate your entire HACCP plan.

Ongoing Verification

Verifi cation is "the use of methods, procedures , or tests in addition to those used in monitoring, to determine whether the HACCP system is operating as intended." After a HACCP plan has been initially validated and put into action, verifi cation activities continue on an ongoing basis.

Simply stated, you need to verify that your HACCP system is working the way you expected.

There are several ways to do this, here are a few: (these aren't the only ones)

- Calibrate your monitoring equipment.
- Sample your product.
- Review your monitoring and corrective action logs.
- Personally inspect your facility's operations.

Whatever types of ongoing verification activities you decide to use, they should be included in your HACCP plan along with the specifics on your CCP's, critical limits, monitoring, and corrective actions. Also, the HACCP team needs to identify the schedules for conducting the verifi cation checks.

Reassessment of the HACCP Plan

It is a good idea to reassess the adequacy of your plan at least once a year and whenever any new changes occur that could affect the hazard analysis or alter the HACCP plan. Here are a few, but not all, of the changes that would require modifi cation to your HACCP plan.

1. Potential new hazards are identified that may be introduced into the process.
2. New ingredients are added, or when an ingredient supplier is changed.
3. The process steps or procedures are changed.
4. New or different processing equipment is introduced.
5. Production volume changes.
6. Personnel changes.

Your reassessment should include a review of the existing HACCP plan, including the product evaluation, hazard analysis, critical control points, critical limits, monitoring procedures, corrective actions and recordkeeping procedures.

Working with the "Verification Procedures" Form

It's important to remember that verification procedures are ongoing activities. For each CCP you will need a monitoring log, a deviation/corrective action log, and an equipment calibration log. These logs are the continual verifi cation that HACCP is being done effectively.

(Like the monitoring form in principle 4, the information on this form is the "Who, What, When and How" of verification.)

For the Example Facility:

- The Who is the quality control supervisor.
- The What is each one of the three activities they need for their process,
- The When is specifi ed after each activity, and
- The How would be determined as needed by the quality control supervisor.

Finishing Your HACCP Plan

Each form that is used in the development of the HACCP plan and the HACCP plan itself needs to be reviewed in its entirety and signed and dated by the responsible offi cial on the HACCP team. This person must make sure that the HACCP plan is complete. This assures the HACCP team that only the most complete and up-to-date plan is being used.

The HACCP System

The HACCP Plan is a written document that is based on the 7 principles of HACCP. A HACCP System is the results of the implementation of the HACCP plan. It includes the written HACCP plan itself but also any records produced, verification data and any prerequisite programs (either written plans or records for GMPs and SSOPs)

The HACCP system produces real results. HACCP is a way of getting and keeping control over your entire production process.

HACCP Principle 7
Verification Procedures Form

Product/Process Name: *Beef Jerky/Heat Treated, Shelf Stable*

Process Step/CCP: *Cooking* *CCP # 01B*

Verification Procedures - (Who, What, When, How)

Quality Control Supervisor

- *Thermometer calibration - Weekly*
- *Random observation of monitoring - Daily*
- *Review relevant records - Daily, prior to shipment*
- *Deviation response review - Ongoing*

Defveloped by: Barbara Jones Date: 21/02/21

Sample Forms

HACCP Team

Store Name ______________________________

Street Address ______________________________

City ______________ State ________ Zip Code ____________

Team Members	Role

Developed by: ______________________ Date ______________

Product/Process Covered

Store Name ____

Street Address ____

City ______________________ State__________ Zip Code ______________

Product/Process Covered Under the HACCP Plan

Smoking/Curing

____
____
____
____
____

Reduced Oxygen Packaging

____
____
____
____
____

Food Additives

____
____
____
____
____

Variances

____
____
____
____
____

Developed by: ______________________ Date ______________________

Ingredients and Raw Materials

Store Name ______________________________

Street Address ______________________________

City ______________ State__________ Zip Code ______________

Product/Process Category______________________________

Product Examples ______________________________

Meat Poultry and Byproducts	Nonmeat Food Ingredients	Binders/Extenders
Spices/Flavorings	Restricted Ingredients	Preservatives/Acidifiers
Liquid	Packaging Materials	Other

Developed by: ______________________ Date______________

Identifiying Critical Control Points

Store Name ____

Street Address ____

City ____________________ State __________ Zip Code ____________

Process/Step ____

Critical Control Point Decision Tree

Question 1A
Do preventative measures exist for the identified hazards?
If no - go to Question 1B.
If yes - go to Question 2.

Question 1B
Is control at this step necessary for safety?
If no - not a CCP
If yes - modify step, process or product and return to Question 1.

Question 2
Does this step eliminate or reduce the likely occurence of a hazard(s) to an acceptable level?
If no - go to Question 3.
If yes - CCP.

Question 3
Could contamination with identified hazard(s) occur in excess of acceptable levels or could these increase to unacceptable levels?
If no - not a CCP.
If yes - go to Question 4.

Question 4
Will a subsequent step eliminate the identified hazards or reduce the likely occurence to an unacceptable level?
If no - CCP.
If yes - not a CCP.

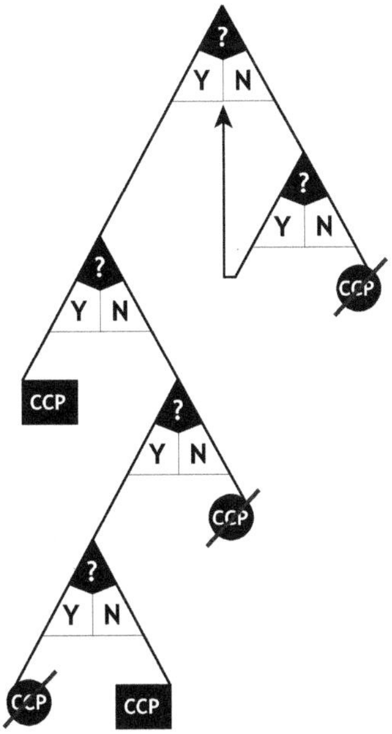

BIOLOGICAL	CHEMICAL	PHYSICAL
❑ CCP#__________ ❑ Not a CCP	❑ CCP#__________ ❑ Not a CCP	❑ CCP#__________ ❑ Not a CCP

Developed by: ____________________ Date ____________

Critical Limits

Store Name ________________________________

Street Address ________________________________

City ________________ State________ Zip Code ____________

Product/Process Name ________________________________

Process Step/CCP ________________________________

CRITICAL LIMITS

Limit *(time, temp, pH, etc.)* - ________________________________

Source *(cite a regulation, scientific document, other resource)* - ________________

Developed by: ________________________ Date________________

Monitoring Procedures

Store Name ______________________________

Street Address ______________________________

City ____________________ State__________ Zip Code ______________

Product/Process Name ______________________________

Process Step/CCP ______________________________

MONITORING PROCEDURES

(Who, What, When, How) - ______________________________

Developed by: ____________________ Date ______________

Corrective Action Procedures

Store Name ______________________________

Street Address ______________________________

City ____________________ State __________ Zip Code ____________

Product/Process Name ______________________________

Process Step/CCP ______________________________

Problem (critical limit exceeded) - ______________________________

____

____

Disposition of Product (hold, rework, condemn) - ______________________________

____

____

____

____

Corrective Action Procedure/Steps - ______________________________

____

____

____

____

Who is responsible for performing these corrective actions?- ______________________________

____

____

____

____

Compliance Procedures- ______________________________

____

____

____

____

Developed by: ____________________ Date ____________________

Hazard Analysis Form

Store Name ______________________________

Street Address ______________________________

City ____________________ State__________ Zip Code ____________

Product/Process Name: ______________________________

Process Step from Flow Diagram: ______________________________

C: CHEMICAL	B: BIOLOGICAL	P: PHYSICAL
List the Hazards:		

Is the hazard reasonably likely to occur?

C: CHEMICAL	B: BIOLOGICAL	P: PHYSICAL
❑ Yes ❑ No	❑ Yes ❑ No	❑ Yes ❑ No

What is the basis for your decision?

What preventative measures can be applied at this step to prevent, eliminate or reduce the hazard to an acceptable level?

____

____

____

Developed by: ______________________________ ***Date*** ____________

Hazard Analysis Worksheet

Store Name ____

Street Address ____

City ______________________ State__________ Zip Code ______________

(1) Ingredient/ processing step	(2) Identify potential hazards introdced, controlled or enhanced at this time	(3) Are any potential food safety hazards significant? (YES/NO)	(4) Justify your decision for column 3	(5) What preventative measure(s) can be applied to prevent the significant hazards?	(6) Is this step a critical control point? (YES/NO)
	BIOLOGICAL CHEMICAL PHYSICAL				
	BIOLOGICAL CHEMICAL PHYSICAL				
	BIOLOGICAL CHEMICAL PHYSICAL				
	BIOLOGICAL CHEMICAL PHYSICAL				
	BIOLOGICAL CHEMICAL PHYSICAL				
	BIOLOGICAL CHEMICAL PHYSICAL				

Developed by: ______________________________ Date__________________

HACCP Plan

Store Name ____

Street Address ____

City ____________________ State__________ Zip Code ______________

Product/Process ________________________ Date ______________

(1) Critical Control Poin (CCP)	(2) Significant Hazards	(3) Critical Limits for each Preventative Measure	(4) Monitoring	(5)	(6)	(7)	(8) Corrective Action(s)	(9) Records	(10) Verification
			What	How	Freq-uency	Who			

Common Foodborne

Bacterial Pathogens

Common Foodborne Bacterial Pathogens

Bacillus cereus

Bacillus cereus is an aerobic spore farmer. Two types of toxins can be produced, one results in diarrheal syndrome and the other in emetic syndrome.

RESERVOIR . . WIDELY DISTRIBUTED IN THE ENVIRONMENT.

IMPLICATED FOODS. . . . . . RICE, MEATS, DAIRY PRODUCTS, VEGETABLES, FISH, PASTA, SAUCES, PUDDINGS, SOUPS, PASTRIES AND SALADS.

B. cereus is widely distributed throughout the environment. It has been isolated from a variety of foods, meats, dairy products, vegetables, fish and rice. The bacteria can be found in starchy foods such as potato, pasta and cheese products, and food mixtures such as sauces, puddings, soups, casseroles, pastries and salads.

GROWTH REQUIREMENTS

TEMPERATURE (F)	39-131
MINIMUM WATER ACTIVITY	0.92
PH	4.3-9.3
MAXIMUM SALT (%)	18
ATMOSPHERE	AEROBE
SURVIVAL CONDITIONS	SALT-TOLERANT, SPORES ARE HEAT RESISTANT

This organism will grow at temperatures as low as 39°F, at a pH as low as 4.3, and at salt concentrations as high as 18%. Unlike other pathogens, it is an aerobe, and will grow only in the presence of oxygen. Both the spores and the emetic toxin are heat-resistant.

CONTROLS. . . . . REFRIGERATION
CONTROL OF *BACILLUS CEREUS* CAN BE ACHIEVED THROUGH PROPER REFRIGERATION.

Campylobacter

Campylobacter jejuni infection, called Campylobacteriosis, causes diarrhea, which may be watery or sticky and maintain blood. Estimated numbers of cases of campylobacteriosis exceed 2-4 million per year, is considered the leading cause of human diarrheal illness in the United States, and is reported to cause more disease than *Shigella* and *Salmonella* spp. combined.

RESERVOIR . . . CHICKENS, COWS, FLIES, CATS, PUPPIES

IMPLICATED FOODS. . . . . . RAW OR UNDERCOOKED CHICKEN, MEAT, SEAFOOD, CLAMS, MILK, EGGS, NON-CHLORINATED WATER, RECONTAMINATED READY-TO-EAT FOODS.

Raw and undercooked chicken, raw and improperly pasteurized milk, raw clams, and non-chlorinated water have been implicated in campylobacteriosis The organism has been isolated from crabmeat. It's carried by healthy chickens and cows, and can be isolated from flies, cats and puppies.

GROWTH REQUIREMENTS

TEMPERATURE (F)	86-113
MINIMUM WATER ACTIVITY	0.99
PH	4.39-9.5
MAXIMUM SALT (%)	1.5
ATMOSPHERE	MICROAEROPHILIC
SURVIVAL CONDITIONS	SENSITIVE TO DRYING, HEATING, DISINFECTION, ACID, AIR

The thing that makes "Campy" unique is its very special oxygen requirements. It's micro-aerophilic, which means it requires reduced levels of oxygen to grow: about 3-15% oxygen (conditions similar to the intestinal tract). Another point worth noting is that it will not grow at temperatures below 86°F, or at salt levels above 1.5%. The organism is considered fragile and sensitive to environmental stresses like drying, heating, disinfection, acid and air which is 21% oxygen. It requires a high water activity and fairly neutral pH for

CONTROLS: . . . . SANITATION TO PREVENT RECONTAMINATION; COOKING; PASTEURIZATION; WATER TREATMENT.

The controls are very basic: proper cooking and pasteurization, proper hygienic practices by food handlers to prevent recontamination, and adequate water treatment.

Clostridium botulinum

Clostridium botulinum is an anaerobic spore-former. Actually there are seven types of Clostridium botulinum - A, B, C, D, E, F and G - but the only ones we'll discuss here are type A, which represents a group of proteolytic bot, type E, which represents the nonproteolytic group. The reason for the distinction is in the proteolytic organisms' ability to break down protein.

This organism is one of the most lethal pathogens covered here. Symptoms include weakness and vertigo, followed by double vision and progressive difficulty in speaking, breathing and swallowing. There may also be abdominal distention and constipation. The toxin eventually causes paralysis, which progresses symmetrically downward, starting with the eyes and face, and proceeding to the throat, chest, and extremities. When the diaphragm and chest muscles become involved, respiration is inhibited, and death from asphyxia results. Treatment includes early administration of antitoxin and mechanical breathing assistance. Mortality is high - without the antitoxin, death is almost certain.

RESERVOIR . . .SOIL;FRESH WATER AND MARINE SEDIMENTS; FISH; MAMMALS

IMPLICATED FOODS: . . . . . .CANNED FOODS; ACIDIFIED FOODS; SMOKED AND UNEVISCERATED FISH; STUFFED EGGPLANT; GARLIC IN OIL; BAKED POTATOES; SAUTEED ONIONS; BLACK BEAN DIP; MEAT PRODUCTS; MARSCAPONE CHEESE.

Bot is widely distributed in nature and can be found in soils, sediments from streams, lakes and coastal waters, the intestinal tracts of fish and mammals, and the gills and viscera of crabs and other shellfish. Type E is most prevalent in fresh water and marine environments, while Type A is generally found terrestrially.

Bot has been a problem in a wide variety of food products: canned foods, acidified foods, smoked and uneviscerated fish, stuffed eggplant, garlic in oil, baked potatoes, sauteed onions, black bean dip, meat products, and marscapone cheese, to name just a few.

Two outbreaks in the 1960's involved vacuum-packaged fish (smoked ciscos and smoked chubs). The causative agent in each case was *C botulinum* type E. The products were packed without nitrates, with low levels of salt, and were temperature-abused during distribution, all of which contributed to the formation of the toxin. There were no obvious signs of spoilage because aerobic spoilage organisms were inhibited by the vacuum packaging, and because type E does not produce any offensive odors.

Three cases of botulism in NY were traced to chopped garlic bottled in oil, which had been held at room temperature for several months before it was opened. Presumably, the oil created an anaerobic environment.

GROWTH REQUIREMENTS

	TYPE A	TYPE E
TEMPERATURE (F)	50-113	38-113
MINIMUM WATER ACTIVITY	0.94	0.97
PH	4.6-9.0	5.0-9.0
MAXIMUM SALT (%)	10	5
ATMOSPHERE	ANAEROBE	
SURVIVAL CONDITIONS	HEAT RESISTANT	

Type A and type E vary in their growth requirements. Minimum growth temperature for type A is 50°F, while type E will tolerate conditions down to 38°F. Type A's minimum water activity is 0.94, and type E's is 0.97 - a small difference on paper, but important in controlling an organism. The acid-tolerance of type A is reached at a pH of 4.6, while type E can grow at a pH of 5. A type A is more salt-tolerant; it can handle up to 10%, when 5% is sufficient to stop the growth of type E.

Although the vegetative cells are susceptible to heat, the spores are heat resistant and able to survive many adverse environmental conditions. Type A and type E differ in the heat-resistance of their spores; compared to E, type A's resistance is relatively high. By contrast, the neurotoxin produced by *C.bot* is not resistant to heat, and can be inactivated by heating for 10 minutes at 176°F.

CONTROLS. . . . . DESTRUCTION: THERMAL PROCESSING

PREVENTION OF TOXIN FORMATION. . ACIDIFICATION, SALT, WATER ACTIVITY CONTROL, NITRITES, REFRIGERATION

There are two primary strategies to control *C. bot*. The first is destruction of the spores by heat (thermal processing). The second is to alter the food to inhibit toxin production - something which can be achieved by acidification, controlling water activity, the use of salt and preservatives, and refrigeration. Water activity, salt and pH can each be individually considered a full barrier to growth, but very often these single barriers - a pH of 4.6 or 10% salt - are not used because they result in a product which is unacceptable to consumers. For this reason multiple barriers are used.

One example of a product using multiple barriers is pasteurized crabmeat stored under refrigeration; here, type E is destroyed by the pasteurization process, while type A is controlled by the refrigerated storage. (Remember that type E is more sensitive to heat, while type A's minimum growth temperature is 50°F.)

Another example of multiple barriers is hot-smoked, vacuum packaged fish. Vacuum packaging provides the anaerobic environment necessary for the growth of *C. bot*, even as it inhibits the normal aerobic spoilage flora which would otherwise offer competition and give telltale signs of spoilage. So heat is used to weaken the spores of type E, which are then further controlled by the use of salt, sometimes in combination with nitrites. Finally spores of type A are controlled by refrigeration.

Vacuum-packaging of foods which are minimally processed, like sous vide products, allows the survival of *C. bot* spores while completely wiping out competing microflora. If no control barriers are present, the *C. bot* may grow and produce toxin, particularly if there is temperature abuse. Given the frequency of temperature abuse documented at the retail and consumer levels, this process is safe only if temperatures are carefully controlled to below 38°F throughout distribution. Vacuum-packaging is also used to extend the shelf-life of the product. Since this provides additional time for toxin development, such food must be considered a high risk.

Clostridium perfringens

Clostridium perfringens is an anaerobic spore former and one of the most common agents of foodborne gastroenteritis. Perfringens poisoning, the disease caused by the organism, is characterized by intense abdominal cramps and diarrhea.

RESERVOIR . . .HUMANS, DOMESTIC AND WILD ANIMALS, SOIL, SEDIMENT

IMPLICATED FOODS. . . . . . .MEAT, POULTRY, GRAVY, CASSEROLES

C. perfringens is widely distributed in the environment and is frequently in the intestines of humans and many domestic and wild animals. Spores of the organism persist in soil and sediments.

C. perfringens has been found in beef, pork, lamb, chicken, turkey, stews, casseroles, and gravy.

GROWTH REQUIREMENTS

TEMPERATURE (F)	50-125
MINIMUM WATER ACTIVITY	0.93
PH	5.0-9.0
MAXIMUM SALT (%)	7
ATMOSPHERE	ANAEROBIC
SURVIVAL CONDITIONS	HEAT-RESISTANT

Clostridium perfringens is a mesophilic organism. Since it is also a spore-former, it is quite resistant to heat, and temperatures for growth range from 50°F to 125°F. pH, water activity and salt ranges for growth are fairly typical.

CONTROLS. . . .PROPER COOLING, HOLDING, AND REHEATING: EDUCATION OF FOOD HANDLERS.

Far from killing the spores, cooking encourages them to germinate when the product reaches a suitable temperature. Rapid, uniform cooling after cooking is needed. In virtually all outbreaks, the principal cause of perfringens poisoning is failure to properly refrigerate previously cooked foods, especially when prepared in large portions. Proper hot holding (above 140°F) and adequate reheating of cooked, chilled foods (to a minimum internal temperature of 165°F) are also necessary controls. The education of food handlers remains the critical aspect of control.

Escherichia coli

There are four classes of pathogenic *E. coli*; enteropathogenic (EPEC), enterotoxigenic (ETEC), enteroinvasive (EIEC), and enterohemmoragic (EHEC). All four types have been associated with food and water borne diseases.

EPEC - Gastroenteritis/infantile diarrhea - Outbreaks have been primarily associated with infants in day-care and nursery settings.

ETCA - Traveler's diarrhea - Contamination of water supplies or food does occasionally lead to outbreaks. Outbreaks have been associated with water and can be contaminated by raw sewage and on imported cheese.

EIEC - Bacillary dysentery - Contaminated water supplies can directly or indirectly (by contaminating food supplies) be the cause of outbreaks; infected food handlers can also be a source.

EHEC - Hemorrhagic colitis - All people are believed to be susceptible to hemorrhagic colitis. The strain E. coli 0157:H7 has become infamous following several outbreaks and probably countless more unreported illnesses. Foods commonly associated with illnesses are undercooked ground beef, unpasteurized apple cider, raw milk, fermented sausage, water and raw vegetables.

GROWTH REQUIREMENTS

TEMPERATURE (F)	45-121
MINIMUM WATER ACTIVITY	0.95
PH	4.0-9.0
MAXIMUM SALT (%)	6.5
ATMOSPHERE	FACULATIVE ANAEROBICE
SURVIVAL CONDITIONS	WITHSTANDS FREEZING AND ACID ENVIRONMENTS

E. coli are mesophilic organisms; they grow best at moderate temperatures, at moderate pH, and in conditions of high water activity. It has, however, been shown that some E. coli strains are very tolerant of acidic environments and freezing.

CONTROLS. . . .PROPER COOKING; PROPER HOLDING TEMPERATURES; PERSONAL HYGIENE; EDUCATION; PREVENTING FECAL CONTAMINATION OF ANIMAL CARCASSES.

Food may be contaminated by infected food handlers who practice poor personal hygiene, or by contact with water contaminated by human sewage. Control measures to prevent food poisoning therefore include educating food workers in safe food handling techniques and proper personal hygiene, properly heated foods, and holding foods under appropriate temperature controls. Additionally, untreated human sewage should not be used to fertilize vegetables and crops used for human consumption, nor should unchlorinated water be used for cleaning food or food contact surfaces.

Prevention of fecal contamination during the slaughter and processing of foods of animal origin is paramount to control foodborne infection of EHEC. Foods of animal origin should be heated sufficiently to kill the organism. Consumers should avoid eating raw or partially cooked meats and poultry, and drinking unpasteurized milk or fruit juices.

Listeria

Listeriosis, the disease caused by this organism, can produce mild flu-like symptoms in healthy individuals. In susceptible individuals, including pregnant women, newborns, and the immunocompromised, the organism may enter the blood stream, resulting in septicemia. Ultimately listeriosis can result in meningitis, encephalitis, spontaneous abortion and still birth.

RESERVOIR . . .SOIL, SILAGE, OTHER ENVIRONMENTAL SOURCES.

IMPLICATED FOODS. . . . . .DAIRY PRODUCTS, VEGETABLES, MEAT, POULTRY, FISH, COOKED READY-TO-EAT PRODUCTS.

L. monocytogenes can be isolated from soil, silage and other environmental sources. It can also be found in man-made environments such as food processing establishments. Generally speaking, however, the drier the environment, the less likely it is to harbor this organism.

L. mono has been associated with raw or inadequately pasteurized milk, cheeses (especially soft-ripened types), ice cream, raw vegetables, fermented sausages, raw and cooked poultry, raw meats, and raw and smoked fish

L. mono is a psychotropic faculative anaerobe. It can survive some degree of thermal processing, but can also be destroyed by cooking to an internal temperature of 158°F for 2 minutes. It can also grow at refrigerated temperatures below 31°F. Reportedly, it has a doubling time of 1.5 days at 40°F. There is nothing unusual about this organisms pH and water activity range for growth. L. mono is salt-tolerant; it can grow in up to 10% salt, and has been known to survive in 30% salt. It is also nitrite-tolerant.

GROWTH REQUIREMENTS
TEMPERATURE (F). 31-113
MINIMUM WATER ACTIVITY 0.92
PH 4.4-9.4
MAXIMUM SALT (%). 10
ATMOSPHERE . . . . FACULATIVE ANEROBE
SURVIVAL CONDITIONS . . .SALT AND NITRITE TOLERANT

CONTROLS. . . .COOKING, PASTEURIZATION, PREVENTION OF RECONTAMINATION

Prevention of recontamination after cooking is a necessary control; even if the product has received thermal processing adequate to inactivate L. monocytogenes, the widespread nature of the organism provides the opportunity for recontamination. Furthermore, if the heat treatment has destroyed the competing microflora, L. mono might find itself in a suitable environment without competition.

Salmonella

There are four syndromes of human salmonellosis: Salmonella gastroenteritis, Typhoid fever; non-typhoidal Salmonella septicemia and asymptomatic carrier. Salmonella gastroenteritis may be caused by any of the Salmonella species other than Salmonella typhi, and is usually a mild, prolonged diarrhea.

True typhoid fever is caused by infection with Salmonella typhi. While fatality rates may exceed 10% in untreated patients, they are less than 1% in patients who receive proper medical treatment. Survivors may become chronic asymptomatic carriers of Salmonella bacteria. Such asymptomatic carriers show no symptoms of the illness, and yet are capable of passing the organisms to others (the classic example is Typhoid Mary).

Non-typhoidal Salmonella septicemia may result from infection with any of the Salmonella species and can affect virtually all organ systems, sometimes leading to death. Survivors may become chronic asymptomatic carriers of Salmonella bacteria.

RESERVOIR . . .DOMESTICATED ANIMALS AND FECES, WATER, SOIL, INSECTS
IMPLICATED FOODS. . . . . . .RAW MEAT, POULTRY, SEAFOOD, EGGS, DAIRY PRODUCT, YEAST, SAUCES, SALAD DRESSINGS, CAKE MIXES, CREAM FILLED DESSERTS, CONFECTIONERY, ETC.

Salmonella often live in animals - especially poultry and swine - as well as in a number of environmental sources. The organisms have been found in water, soil and insects, on factory and kitchen surfaces, and in animal feces. They can also survive in a variety of foods, including raw meats and poultry, dairy products and eggs, fish, shrimp and frog legs, yeast, coconut, sauces and salad dressing, cake mixes, cream-filled desserts and toppings, dried gelatin, peanut butter, orange juice, cocoa and chocolate.

GROWTH REQUIREMENTS
TEMPERATURE (F). 41-115
MINIMUM WATER ACTIVITY 0.94
PH 3.7-9.5
MAXIMUM SALT (%). 8
ATMOSPHERE . . . . FACULATIVE ANAEROBE
SURVIVAL CONDITIONS . . . SENSITIVE TO MODERATE HEAT

Salmonella spp. are also mesophilic organisms which grow best at moderate temperatures and pH, and under conditions of low salt and of high water activity. They are killed rapidly by moderate heat treatment, yet mild heat treatment may give them the ability to develop heat resistance up to 185°F. Similarly, the organisms can adapt to an acidic environment.

CONTROLS . . . .SANITATION TO PREVENT RECONTAMINATION, COOKING, PASTEURIZATION, PROPER HOLDING TEMPERATURES.

Ordinary household cooking, personal hygiene to prevent recontamination of cooked food, and control of time and temperature are generally adequate to prevent salmonellosis.

Shigela

There are actually four species of Shigella. Because there is little difference in their behavior, however, they will be discussed collectively.

Illness is Shigellosis, typical symptoms include fever, cramps, inflammation and ulceration of intestine, and diarrhea. This disease is easily transmitted from person to person.

RESERVOIR . . .HUMAN, ANIMAL
IMPLICATED FOODS. . . . . . .SALADS, RAW VEGETABLES, POULTRY, MEAT, FISH, FRUIT, DAIRY PRODUCTS, BAKERY PRODUCTS.

The only significant reservoir for Shigella is humans. Foods associated with shigellosis include salads (potato, tuna, shrimp, macaroni and chicken), raw vegetables, milk and dairy products, poultry, fruits, bakery products, hamburger and fin fish.

GROWTH REQUIREMENTS

TEMPERATURE (F)	43-117
MINIMUM WATER ACTIVITY	0.96
PH	4.8-9.3
MAXIMUM SALT (%)	5
ATMOSPHERE	FACULATIVE ANAEROBE
SURVIVAL CONDITIONS	SURVIVES ACIDIC CONDITIONS

The growth conditions for *Shigella*, which are mesophilic organisms, are similar to those of *Salmonella*. *Shigella* can survive under various environmental conditions, including low acid.

CONTROLS . . . .COOKING, PROPER HOLDING TEMPERATURES, SANITATION TO PREVENT RECONTAMINATION, ADEQUATE WATER TREATMENT.

Shigella can spread rapidly under the crowded and unsanitary conditions often found in such places as summer camps, refugee camps and camps for migrant workers, and at mass gatherings such as music festivals. The primary reasons for the spread of Shigella in foods are poor personal hygiene on the part of food handlers, and the use of improper holding temperatures for contaminated foods; conversely, the best preventive measures would be good personal hygiene and health education. Chlorination of water and sanitary disposal of sewage would prevent waterborne outbreaks of shigellosis.

Staphylococcus aureus

Staphylococcus aureus produces a highly heat-stable toxin. Staphylococal food poisoning is one of the most economically important foodborne diseases in the US., costing approximately $1.5 billion each year in medical expenses and loss of productivity. The most common symptoms are nausea, vomiting, abdominal cramps, diarrhea and prostration.

RESERVOIR . . .HUMANS, ANIMALS, AIR, DUST, SEWAGE, WATER

IMPLICATED FOODS. . . . . . .POULTRY, MEAT, SALADS, BAKERY PRODUCTS, SANDWICHES, DAIRY PRODUCTS.

Staph can be found in air, dust, sewage and water, although humans and animals are the primary reservoirs. *Staph* is present in and on the nasal passages, throats, hair and skin of at least one out of two healthy individuals. Food handlers are the main source of contamination, but food equipment and the environment itself can also be sources of the organism.

Foods associated with *Staph* include poultry, meat, salads, bakery products, sandwiches and dairy products. Due to poor hygiene and temperature abuse, a number of outbreaks have been associated with cream-filled pastries and salads such as egg, chicken, tuna, potato, and macaroni.

GROWTH REQUIREMENTS

TEMPERATURE (F)	
GROWTH	45-122
TOXIN PRODUCTION	50-118
MINIMUM WATER ACTIVITY	
GROWTH	0.83
TOXIN PRODUCTION	0.85
PH	4.0-10.0
MAXIMUM SALT (%)	
GROWTH	25
TOXIN PRODUCTION	10
ATMOSPHERE	FACULATIVE ANAEROBIC
SURVIVAL CONDITIONS	TOLERANT OF HIGH SALT AND LOW MOISTURE

S. aureus grows and produces toxin at the lowest water activity (0.85) of any food pathogen. And, like type *A bot* and *Listeria*, *Staph* is quite salt-tolerant and will produce toxin at 10%.

CONTROLS. . . .HEATING, PROPER EMPLOYEE HYGIENE, PREVENTION OF TEMPERATURE ABUSE

Foods which require considerable handling during preparation and which are kept at slightly elevated temperatures after preparation are frequently involved in staphylococcal food poisoning. And, while *S. aureus* does not compete well with the bacteria normally found in raw foods, it will grow both in cooked products and in salted products where the salt inhibits spoilage bacteria. Since Staph is a faculative anaerobe, reduced oxygen packaging can also give it a competitive advantage.

The best way to control Staph is to ensure proper employee hygiene and to minimize exposure to uncontrolled temperatures. Remember that while the organism can be killed by heat, the toxin cannot be destroyed even by heating.

Vibrios

There are quite a few species of Vibrios, but only four will be covered.

Vibrio parahaemolyticus - The bacteria is naturally occurring in estuaries and other coastal waters. Illness is most commonly associated with fish and shellfish which are raw, undercooked or recontaminated after cooking.

Vibrio cholerae 01 - Epidemic cholera - Poor sanitation and contaminated water supplies will spread the disease; feces contaminated foods including seafood have also been associated with outbreaks.

Vibrio cholerae non-01 - The reservoir for this organism is estuarine water - illness is associated with raw oysters, but the bacteria has also been found in crabs.

Vibrio vulnificus - This organism also occurs naturally in estuarine waters. So far only oysters from the Gulf of Mexico have been implicated in illness, but the organism itself has been found in both the Atlantic and Pacific Oceans.

GROWTH REQUIREMENTS

TEMPERATURE (F). 41-111
MINIMUM WATER ACTIVITY 0.94-0.97
PH 4.8-11.0
MAXIMUM SALT (%). 5-10
ATMOSPHERE . . . . FACULATIVE ANAEROBE
SURVIVAL CONDITIONS . . . SALT TOLERANT; HEAT SENSITIVE

Vibrios are mesophilic and require relatively warm temperatures, high water activity and come neutral pH for growth, they also require some salt for growth, and are quite salt-tolerant. They are, however, easily eliminated by a mild heat treatment.

CONTROLS. . . .COOKING, PREVENTION OF RECONTAMINATION, TIME/TEMPERATURE ABUSE, CONTROL PRODUCT SOURCE.

All the Vibrios can be controlled through cooking and the prevention of cross-contamination afterward. Proper refrigeration prevents proliferation, which is particularly important because of the short generation times for these species. To guard against cholerae, processors should know the source of the product and be cautious about importing from countries experiencing an epidemic.

Yersinia

Yersinia ssp: *Y. entercolitica*; Y. *pseudotuberculosis*; *Y. pestis*
Of the 11 recognized species of *Yersinia*, three are know to be potentially pathogenic to humans: enterocolitica, pseudotuberculosis and pestis. Only enterocolitica and pseudotuberculosis are recognized as foodborne pathogens. *Y. pestis*, the organism responsible for the black plague, is not transmitted by food.

Yersiniosos is often characterized by such symptoms as gastroenteritis with diarrhea and/or vomiting, but fever and abdominal pain are the hallmark symptoms. Yersinia infections mimic appendicitis, which has led to unnecessary operations.

RESERVOIR . . .LAKES, STREAMS, VEGETATION,SOIL, BIRDS, ANIMALS AND THEIR FECES

IMPLICATED FOODS. . . . . . .RAW VEGETABLES, MILK, ICE CREAM, CAKE, PORK, SOY, SALAD, SEAFOOD, CLAMS, SHRIMP

Yersinia can be found in raw vegetables, milk, ice cream, cakes, pork, soy products, salads, oysters, clams and shrimp. They are found in the environment, in such places as lakes, streams, soil and vegetation. They've been isolated from the feces of dogs, cats, goats, cattle, chincillas, mink, and primates; in the estuarine environment, many birds - among them, waterfowl and seagulls - may be carriers. The foodborne nature of Yersiniosis is well established, and numerous outbreaks have occurred worldwide.

GROWTH REQUIREMENTS

TEMPERATURE (F). 30-108
MINIMUM WATER ACTIVITY 0.95
PH 4.2-10.0
MAXIMUM SALT (%). 7
ATMOSPHERE . . . . FACULATIVE ANAEROBE
SURVIVAL CONDITIONS . . . WITHSTANDS FREEZING AND THAWING; SENSITIVE TO HEATING AND SANITIZERS

CONTROLS. . . .SANITATION TO PREVENT RECONTAMINATION; COOKING; PASTEURIZATION; WATER TREATMENT; PROPER HOLDING TEMPERATURES

Key factors for controlling Yersinia include proper cooking or pasteurization, proper food handling to prevent recontamination, adequate water treatment, and care taken to ensure that products are not time or temperature abused. Proper use of sanitizers is also an effective control. Essentially, to control Yersinia, it is necessary to keep things clean and moving.

Bibliography:

1. Tauxe R.V. Emerging foodborne diseases: an evolving public health challenge. Emerging Infectious Diseases1997; 3 (4): 425-434.

2. Buzby J.C. Older adults at risk of complications from microbial foodborne illness. Food Rev 2002; 25 (1): 30-35

3. Rocourt J, Moy G, Vierk K, Schlundt, J. (2003). The Present State of Foodborne Disease in OECD Countries (on-line). Food Safety Department, WHO, Geneva, Switzerland, Available from: http://www.who.int/foodsafety/publications/foodborne_disease/oesd_fbd.pdf.

4. Lund M. B, Hunter R. P. The Microbiological Safety of Food in Healthcare Settings, 2008; Blackwell Publishing.

5. Worsfold D. A guide to HACCP and function catering. J. R. Soc. Health 2001; 121(4):224-9.

6. University of California, San Francisco Children's Hospital (UCSF) Bone Marrow Transplant. Dietary Concerns During Bone Barrow Transplant, 2006; Available from: http://www.ucsfhealth.org/childrens /medical_services/cancer/btm/diet.html

7. Dryden M.S, Keyworh N, Gabb R, Stein K. Asymptomatic food handlers as the source of nosocomial salmonellosis. J Hosp Infect 1994; 28 (3): 195-208.

8. Richards J, Parr E, Riseborough P. Hospital food hygiene: The application of Hazard Analysis Critical Control Points to conventional hospital catering. Journal of Hospital Infectio1993; 24 (4):273 – 278.

9. Wall P.G, Ryan M.J, Ward L.R, Rowe B. Outbreaks of Salmonellossis in hospitals in England and Wales: 1992-1994. Journal of Hospital Infection 1996; 33 (3):181-190.

10. Sharp J.C, Collier P.W, Gilbert R.J. Food poisoning in hospitals in Scotland. Journal of Hygiene (Lond)1979; 83 (2): 231-236.

11. Dalton C.B, Gregory J, Kirk M.D. Foodborne outbreaks in Australia, 1995 – 2000. Communicable Disease Intelligence 2004; 28 (2): 211-224.

12. Abbott J.D, Hepner E.D, Clifford C. Salmonella infection in hospital. A report from the Public Health Laboratory Service Salmonella Subcommittee. Journal of Hospital Infection 1980; 1 (4):307-314.

13. Welinder-Olsson C, Stenqvist K, Badenfors M, Brandberg A, Florén K, Holm M, Holmberg L, Kjellin E, Mårild S, Studahl A, Kaijser B. EHEC outbreak among staff at a

children's hospital – use of PCR for verocytotoxin detection and PFGE for epidemiological investigation. Epidemiology and Infection 2004; 132 (1): 43049.

14. Mason B.W, Williams N, Salmon R.L, Lewis A, Price J, Johnstan K.M, Trott R.M. Outbreak of Salmonella Indiana associated with egg mayonnaise sandwiches at an acute NHS hospital. Communicable Disease and Public Health 2001; 4 (4): 300-304.

15. Haeghebaert S, Duché L, Gilles C, Masini B, Dubreuil M, Minet J.C, Bouvet P, Grimont F, Delarocque Astagneau E, Vaillant V. Minced beef and human salmonellosis: review of the investigation of three outbreaks in France. Eurosurveillance Monthly 2001; 6(2): 21-26.

16. Gikas A, Kritsotakis E.I, Maraki S, Roumbelaki M, Babalis D, Scoulica E, Panoulis C, Saloustros E, Kontopodis E, Samonis G, Tselentis Y. A nosocomial, foodborne outbreak of Salmonella Enteritica serovar Enteritidis in a university hospital in Greece: the importance of establishing HACCP systems in hospital catering. J Hosp Infect 2007; 66 (2):194-6.

17. Van Acker J, De Smet F, Muyldermans G, Bouqatef A, Naessem A, Lauwers S. Outbreak of necrotizing enterocolitis associated with Enterobacter sakazakkii in powdered milk formula. Journal of Clinical Microbiology 2001; 39 (1): 293-297.

18. Lo S.V, Connolly A.M, Palmer S.R, Wright D, Thomas P.D, Joynson D. The role of the pre-symptomatic food handler in a common source outbreak of food-borne SRSV gastroenteritis in a group of hospitals. Epidemiology and Infection1994; 113 (3): 513-521.

19. Bauman H.E. The HACCP concept and microbiological hazard categories. Food Technolog 1974; 28 (1): 30-4, 74.

20. Codex Committee on Food Hygiene (1993) Guidelines for the Application of the Hazard Analysis Critical Control Point (HACCP) System, in Training Considerations for the Application of the HACCP System to Food Processing and Manufacturing, WHO/FNU/FOS/93.3 II, World Health Organization, Geneva.

21. Directive 93/43/EEC of 14 June 1993 on the hygiene of foodstuffs, OJ L175, 1993.

22. EC (2004) Regulation (EC) No 852/2004 of the European Parliament and of the Council of 29 April 2004 on the hygiene of foodstuffs. Official Journal of the European Union L139/30.04.04.

23. Codex Committee on Food Hygiene (1997a) Recommended International Code of Practice, General Principles of Food Hygiene, CAC/RCP 1-1969, Rev. 3 (1997) in Codex Alimentarius Commission Food Hygiene Basic Texts, Food and Agriculture Organi;zation of the United Nations, World Health Organization, Rome.

24. NACMCF - National Advisory Committee on Microbiological Criteria for Foods, Hazard Analysis and Critical Control Points Principles and Application Guidelines, 1997.

25. Mortimore S, Wallace C. HACCP. Food Industry Briefing Series 2001; Blackwell Publishing.
26. FDA/CFSAN (2004) FDA Report on the Occurrence of Foodborne Illness Risk Factors in Selected Institutional Foodservice, Restaurant, and Retail Food Store Facility Types. Available from http://www.cfsan.fda.gov/~acrobat/retrsk2.pdf. Accessed 5 March 2007.
27. Anderson A. Enteral tube feeds as a source of infection: can we reduce the risk? Nutrition 1999; 15 (1): 55-57.
28. Oliveira M.R, Batista C.R, Aidoo KE. Application of Hazard Analysis Critical Control Points system to enteral tube feeding in hospital. J Hum Nutr Diet 2001; 14 (5): 397-403.
29. Patchell C, Anderson A, Holden C, MacDonald A, George R.H, Booth I.W. Reducing bacterial contamination of enteral feeds. Arch. Dis. Child 1998; 78 (1):166-168.
30. Buccheri C, Casuccio A, Giammanco S, Giammanco M, La Guardia M, Mammina C. Food Safety in hospital: knowledge, attitudes and practices of nursing staff of two hospitals in Sicily, Italy. BMC Health Serv Res 2007;3 (7):45.
31. McCall B, McCormack J.C, Stafford R, Towner C. An outbreak of Salmonella typhimuriuim at a Teaching Hospital Infection. Control and Hospital Epidemiology 1999; 20 (1): 55-56.
32.HACCP plan, Minnesota Department of Agriculture, Dairy and Food Inspection Division, Minnesota, USA
33. Spearing N.M, Jensen A, McCall B.J, Neill A.S, McCormack G.J. Direct costs associated with a nosocomial outbreak of Salmonella infection: an ounce of prevention is worth a pound of cure. Am J Infect Control 2000; 28 (1): 54-57.
34. Murphy O, Gray J, Gordon S, Bint A. J. An outbreak of Campylobacter food poisoning in a healthcare setting. Journal of Hospital Infection1995; 30 (3): 225-228.
35. Conway C. Food Safety. Nurs Stand 2001;15 (41):47-52.
36. Angelillo F. I, Viggiani M.A.N, Greco M. R, Rito D. HACCP and food hygiene in hospitals: knowledge, attitudes, and practices of food – service staff in Calabria, Italy. Infection control and hospital epidemiology 2001; 22 (6): 363-369.
37.0. Baird D.R, Henry M, Liddell K.G, Mitchell C.M, Sneddon J.G. Post- operative endophthalmitis: the application of hazard analysis critical control points (HACCP) to an infection control problem. Journal of Hospital Infection 2001;49(1):14-22.

	**Vol.1** For All Food Industries (Meat, Beverage, Dairy, Bakeries, Tortilla, Grain and Oilseed, Fruit and Vegetable, Seafood, And Sugar and Confectionery)
	**Vol.2** For All Energy & Electrical Industries (Renewable Energy, Biofuels, Solar Heating & Cooling, Hydroelectric Power, Solar Power, Wind Power, Energy Conservation, Geothermal and Nuclear Power)
	**Vol.3** For All Fashion & Textile Industries (Fashion Design, The Fashion System, Fashion Retailing, Marketing and Marchandizing, Textile Design and Production, Clothing and Textile Recycling)
	**Vol.4** For All Health & Beauty Industries (Fragrances, Makeup, Cosmetics, Personal Care, Sunscreen, Toothpaste, Bathing, Nailcare & Shaving, Skin Care, Foot Care, Hair Care and Other Health & Beauty Products)

<table>
<tr><td>
</td><td>Vol.5
For All Maintenance & Cleaning Products (All-purpose Cleaners, Abrasive Cleaners, Powders. Liquids, Specialty Cleaners, Kitchen, Bathroom, Glass and Metal Cleaners, Bleaches, Disinfectants and Disinfectant Cleaners)</td></tr>
<tr><td>
</td><td>Vol.6
For All Wood & Stationery Industries (Wooden Products, Cardboard, Papers, Markers, Pens, NoteBooks. Writing Pads and Writing Sets, Pencils, White Papers, Envelopes and Organizers, Staplers and Paper Clips)</td></tr>
<tr><td>
</td><td>Vol.7
For All DIY & Construction Industries (Do it yourself" ("DIY") of Building, Modifying, or Repairing, Renovation, Construction Materials, Cement, Coarse Aggregates. Clay Bricks, Power Cables, Pipes and Fittings, Plywood, Tiles, Natural Flooring)</td></tr>
<tr><td>
</td><td>Vol.8
For All Agricuture & Gardening Industries (Shifting Cultivation, Nomadic Herding, Livestock Ranching, Commercial Plantations, Mixed Farming, Horticulture, Butterfly Gardens, Container Gardening, Demonstration Gardens, Organic Gardening)</td></tr>
</table>

Vol.9

For All Professional Products & Services (Teachers, Pilots, Lawyers, Advertising Professionals, Architects, Accountants, Engineers, Consultants, Human Resources Specialist, R&D, Psychologists, Pharmacist, Commercial Banker, Research Analyst)

Vol.10

For All Financial Products & Services (Banking, Professional Advisory, Wealth Management, Mutual Funds, Insurance, Stock Market, Treasury/Debt Instruments, Tax/Audit Consulting, Capital Restructuring, Portfolio Management)

Vol.11

For All Tourism Industries (Airline Industry, Travel Agent, Car Rental, Water Transport, Coach Services, Railway, Spacecraft, Hotels, Shared Accommodation, Camping, Bed & Breakfast, Cruises, Tour Operators)

INTERNATIONAL ECO SHOPPING GUIDE

For All Supermarket Customers

CPSIA information can be obtained
at www.ICGtesting.com
Printed in the USA
BVHW041442140321
602473BV00002B/10

9 781777 526856